美国数学会经典影印系列

出版者的话

近年来，我国的科学技术取得了长足进步，特别是在数学等自然科学基础领域不断涌现出一流的研究成果。与此同时，国内的科研队伍与国外的交流合作也越来越密切，越来越多的科研工作者可以熟练地阅读英文文献，并在国际顶级期刊发表英文学术文章，在国外出版社出版英文学术著作。

然而，在国内阅读海外原版英文图书仍不是非常便捷。一方面，这些原版图书主要集中在科技、教育比较发达的大中城市的大型综合图书馆以及科研院所的资料室中，普通读者借阅不甚容易；另一方面，原版书价格昂贵，动辄上百美元，购买也很不方便。这极大地限制了科技工作者对于国外先进科学技术知识的获取，间接阻碍了我国科技的发展。

高等教育出版社本着植根教育、弘扬学术的宗旨服务我国广大科技和教育工作者，同美国数学会（American Mathematical Society）合作，在征求海内外众多专家学者意见的基础上，精选该学会近年出版的数十种专业著作，组织出版了"美国数学会经典影印系列"丛书。美国数学会创建于1888年，是国际上极具影响力的专业学术组织，目前拥有近30000会员和580余个机构成员，出版图书3500多种，冯·诺依曼、莱夫谢茨、陶哲轩等世界级数学大家都是其作者。本影印系列涵盖了代数、几何、分析、方程、拓扑、概率、动力系统等所有主要数学分支以及新近发展的数学主题。

我们希望这套书的出版，能够对国内的科研工作者、教育工作者以及青年学生起到重要的学术引领作用，也希望今后能有更多的海外优秀英文著作被介绍到中国。

高 等 教 育 出 版 社

2016 年 12 月

Higher Order Fourier Analysis

高阶傅里叶分析

Terence Tao

高等教育出版社·北京

To Garth Gaudry, who set me on the road;
To my family, for their constant support;
And to the readers of my blog, for their feedback and contributions.

Contents

Preface		ix
Acknowledgments		x
Chapter 1. Higher order Fourier analysis		1
§1.1.	Equidistribution of polynomial sequences in tori	2
§1.2.	Roth's theorem	26
§1.3.	Linear patterns	45
§1.4.	Equidistribution of polynomials over finite fields	59
§1.5.	The inverse conjecture for the Gowers norm I. The finite field case	74
§1.6.	The inverse conjecture for the Gowers norm II. The integer case	92
§1.7.	Linear equations in primes	109
Chapter 2. Related articles		129
§2.1.	Ultralimit analysis and quantitative algebraic geometry	130
§2.2.	Higher order Hilbert spaces	149
§2.3.	The uncertainty principle	162
Bibliography		179
Index		185

Preface

Traditionally, Fourier analysis has been focused on the analysis of functions in terms of linear phase functions such as the sequence $n \mapsto e(\alpha n) := e^{2\pi i \alpha n}$. In recent years, though, applications have arisen—particularly in connection with problems involving linear patterns such as arithmetic progressions—in which it has been necessary to go beyond the linear phases, replacing them to higher order functions such as quadratic phases $n \mapsto e(\alpha n^2)$. This has given rise to the subject of *quadratic Fourier analysis* and, more generally, to *higher order Fourier analysis*.

The classical results of Weyl on the equidistribution of polynomials (and their generalisations to other orbits on homogeneous spaces) can be interpreted through this perspective as foundational results in this subject. However, the modern theory of higher order Fourier analysis is very recent indeed (and still incomplete to some extent), beginning with the breakthrough work of Gowers [**Go1998**], [**Go2001**] and also heavily influenced by parallel work in ergodic theory, in particular, the seminal work of Host and Kra [**HoKr2005**]. This area was also quickly seen to have much in common with areas of theoretical computer science related to polynomiality testing, and in joint work with Ben Green and Tamar Ziegler [**GrTa2010**], [**GrTa2008c**], [**GrTaZi2010b**], applications of this theory were given to asymptotics for various linear patterns in the prime numbers.

There are already several surveys or texts in the literature (e.g. [**Gr2007**], [**Kr2006**], [**Kr2007**], [**Ho2006**], [**Ta2007**], [**TaVu2006**]) that seek to cover some aspects of these developments. In this text (based on a topics graduate course I taught in the spring of 2010), I attempt to give a broad tour of this nascent field. This text is not intended to directly substitute for the core papers on the subject (many of which are quite technical and lengthy), but focuses instead on basic foundational and preparatory material, and on the simplest illustrative examples of key results, and should thus hopefully serve as a companion to the existing literature on the subject. In accordance with this complementary intention of this text, we also present certain approaches to the material that is not explicitly present in the literature, such as the abstract approach to Gowers-type norms (Section 2.2) or the ultrafilter approach to equidistribution (Section 1.1.3).

There is, however, one important omission in this text that should be pointed out. In order to keep the material here focused, self-contained, and of a reasonable length (in particular, of a length that can be mostly covered in a single graduate course), I have focused on the combinatorial aspects of higher order Fourier analysis, and only very briefly touched upon the equally significant ergodic theory side of the subject. In particular, the breakthrough work of Host and Kra [**HoKr2005**], establishing an ergodic-theoretic precursor to the inverse conjecture for the Gowers norms, is not discussed in detail here; nor is the very recent work of Szegedy [**Sz2009**], [**Sz2009b**], [**Sz2010**], [**Sz2010b**] and Camarena-Szegedy [**CaSz2010**] in which the Host-Kra machinery is adapted to the combinatorial setting. However, some of the foundational material for these papers, such as the ultralimit approach to equidistribution and structural decomposition, or the analysis of parallelopipeds on nilmanifolds, is covered in this text.

This text presumes a graduate-level familiarity with basic real analysis and measure theory, such as is covered in [**Ta2011**], [**Ta2010**], particularly with regard to the "soft" or "qualitative" side of the subject.

The core of the text is Chapter 1, which comprises the main lecture material. The material in Chapter 2 is optional to these lectures, except for the ultrafilter material in Section 2.1 which would be needed to some extent in order to facilitate the ultralimit analysis in Chapter 1. However, it is possible to omit the portions of the text involving ultrafilters and still be able to cover most of the material (though from a narrower set of perspectives).

Acknowledgments

I am greatly indebted to my students of the course on which this text was based, as well as many further commenters on my blog, including Sungjin Kim, William Meyerson, Joel Moreira, Thomas Sauvaget, Siming Tu, and Mads Sørensen. These comments, as well as the original lecture notes for this course, can be viewed online at

terrytao.wordpress.com/category/teaching/254b-higher-order-fourier-analysis/

Thanks also to Ben Green for suggestions. The author is supported by a grant from the MacArthur Foundation, by NSF grant DMS-0649473, and by the NSF Waterman award.

Chapter 1

Higher order Fourier analysis

1.1. Equidistribution of polynomial sequences in tori

(Linear) *Fourier analysis* can be viewed as a tool to study an arbitrary function f on (say) the integers $\mathbf{Z}$, by looking at how such a function correlates with *linear phases* such as $n \mapsto e(\xi n)$, where $e(x) := e^{2\pi i x}$ is the fundamental character, and $\xi \in \mathbf{R}$ is a frequency. These correlations control a number of expressions relating to f, such as the expected behaviour of f on arithmetic progressions $n, n+r, n+2r$ of length three.

In this text we will be studying higher-order correlations, such as the correlation of f with quadratic phases such as $n \mapsto e(\xi n^2)$, as these will control the expected behaviour of f on more complex patterns, such as arithmetic progressions $n, n+r, n+2r, n+3r$ of length four. In order to do this, we must first understand the behaviour of *exponential sums* such as

$$\sum_{n=1}^{N} e(\alpha n^2).$$

Such sums are closely related to the *distribution* of expressions such as αn^2 mod 1 in the unit circle $\mathbf{T} := \mathbf{R}/\mathbf{Z}$, as n varies from 1 to N. More generally, one is interested in the distribution of polynomials $P \colon \mathbf{Z}^d \to \mathbf{T}$ of one or more variables taking values in a torus $\mathbf{T}$; for instance, one might be interested in the distribution of the quadruplet $(\alpha n^2, \alpha(n+r)^2, \alpha(n+2r)^2, \alpha(n+3r)^2)$ as n, r both vary from 1 to N. Roughly speaking, once we understand these types of distributions, then the general machinery of quadratic Fourier analysis will then allow us to understand the distribution of the quadruplet $(f(n), f(n+r), f(n+2r), f(n+3r))$ for more general classes of functions f; this can lead for instance to an understanding of the distribution of arithmetic progressions of length 4 in the primes, if f is somehow related to the primes.

More generally, to find arithmetic progressions such as $n, n+r, n+2r, n+3r$ in a set A, it would suffice to understand the equidistribution of the quadruplet[1] $(1_A(n), 1_A(n+r), 1_A(n+2r), 1_A(n+3r))$ in $\{0,1\}^4$ as n and r vary. This is the starting point for the fundamental connection between *combinatorics* (and more specifically, the task of finding patterns inside sets) and *dynamics* (and more specifically, the theory of equidistribution and recurrence in measure-preserving dynamical systems, which is a subfield of *ergodic theory*). This connection was explored in the previous monograph [**Ta2009**]; it will also be important in this text (particularly as a source of motivation), but the primary focus will be on finitary, and Fourier-based, methods.

---

[1] Here 1_A is the *indicator function* of A, defined by setting $1_A(n)$ equal to 1 when $n \in A$ and equal to zero otherwise.

1.1. Equidistribution in tori

The theory of equidistribution of polynomial orbits was developed in the linear case by Dirichlet and Kronecker, and in the polynomial case by Weyl. There are two regimes of interest; the (qualitative) *asymptotic regime* in which the scale parameter N is sent to infinity, and the (quantitative) *single-scale regime* in which N is kept fixed (but large). Traditionally, it is the asymptotic regime which is studied, which connects the subject to other asymptotic fields of mathematics, such as dynamical systems and ergodic theory. However, for many applications (such as the study of the primes), it is the single-scale regime which is of greater importance. The two regimes are not directly equivalent, but are closely related: the single-scale theory can be usually used to derive analogous results in the asymptotic regime, and conversely the arguments in the asymptotic regime can serve as a simplified model to show the way to proceed in the single-scale regime. The analogy between the two can be made tighter by introducing the (qualitative) *ultralimit regime*, which is formally equivalent to the single-scale regime (except for the fact that explicitly quantitative bounds are abandoned in the ultralimit), but resembles the asymptotic regime quite closely.

For the finitary portion of the text, we will be using *asymptotic notation*: $X \ll Y$, $Y \gg X$, or $X = O(Y)$ denotes the bound $|X| \leq CY$ for some absolute constant C, and if we need C to depend on additional parameters, then we will indicate this by subscripts, e.g., $X \ll_d Y$ means that $|X| \leq C_d Y$ for some C_d depending only on d. In the ultralimit theory we will use an analogue of asymptotic notation, which we will review later in this section.

1.1.1. Asymptotic equidistribution theory. Before we look at the single-scale equidistribution theory (both in its finitary form, and its ultralimit form), we will first study the slightly simpler, and much more classical, *asymptotic* equidistribution theory.

Suppose we have a sequence of points $x(1), x(2), x(3), \ldots$ in a compact metric space X. For any finite $N > 0$, we can define the probability measure

$$\mu_N := \mathbf{E}_{n \in [N]} \delta_{x(n)}$$

which is the average of the *Dirac point masses* on each of the points $x(1), \ldots, x(N)$, where we use $\mathbf{E}_{n \in [N]}$ as shorthand for $\frac{1}{N}\sum_{n=1}^{N}$ (with $[N] := \{1, \ldots, N\}$). *Asymptotic equidistribution theory* is concerned with the limiting behaviour of these probability measures μ_N in the limit $N \to \infty$, for various sequences $x(1), x(2), \ldots$ of interest. In particular, we say that the sequence $x \colon \mathbf{N} \to X$ is *asymptotically equidistributed* on $\mathbf{N}$ with respect to a reference *Borel probability measure* μ on X if the μ_N converge in the vague topology to μ or, in other words, that

$$(1.1) \qquad \mathbf{E}_{n \in [N]} f(x(n)) = \int_X f \, d\mu_N \to \int_X f \, d\mu$$

for all continuous scalar-valued functions $f \in C(X)$. Note (from the *Riesz representation theorem*) that any sequence is asymptotically equidistributed with respect to at most one Borel probability measure μ.

It is also useful to have a slightly stronger notion of equidistribution: we say that a sequence $x\colon \mathbf{N} \to X$ is *totally asymptotically equidistributed* if it is asymptotically equidistributed on every infinite arithmetic progression, i.e. that the sequence $n \mapsto x(qn+r)$ is asymptotically equidistributed for all integers $q \geq 1$ and $r \geq 0$.

A doubly infinite sequence $(x(n))_{n \in \mathbf{Z}}$, indexed by the integers rather than the natural numbers, is said to be asymptotically equidistributed relative to μ if both halves[2] of the sequence $x(1), x(2), x(3), \ldots$ and $x(-1), x(-2), x(-3), \ldots$ are asymptotically equidistributed relative to μ. Similarly, one can define the notion of a doubly infinite sequence being totally asymptotically equidistributed relative to μ.

Example 1.1.1. If $X = \{0,1\}$, and $x(n) := 1$ whenever $2^{2j} \leq n < 2^{2j+1}$ for some natural number j and $x(n) := 0$ otherwise, show that the sequence x is not asymptotically equidistributed with respect to any measure. Thus we see that asymptotic equidistribution requires all scales to behave "the same" in the limit.

Exercise 1.1.1. If $x\colon \mathbf{N} \to X$ is a sequence into a compact metric space X, and μ is a probability measure on X, show that x is asymptotically equidistributed with respect to μ if and only if one has

$$\lim_{N \to \infty} \frac{1}{N} |\{1 \leq n \leq N : x(n) \in U\}| = \mu(U)$$

for all open sets U in X whose boundary ∂U has measure zero. (*Hint:* For the "only if" part, use *Urysohn's lemma*. For the "if" part, reduce (1.1) to functions f taking values between 0 and 1, and observe that almost all of the level sets $\{y \in X : f(y) < t\}$ have a boundary of measure zero.) What happens if the requirement that ∂U have measure zero is omitted?

Exercise 1.1.2. Let x be a sequence in a compact metric space X which is equidistributed relative to some probability measure μ. Show that for any open set U in X with $\mu(U) > 0$, the set $\{n \in \mathbf{N} : x(n) \in U\}$ is infinite, and furthermore has positive lower density in the sense that

$$\liminf_{N \to \infty} \frac{1}{N} |\{1 \leq n \leq N : x(n) \in U\}| > 0.$$

In particular, if the support of μ is equal to X, show that the set $\{x(n) : n \in \mathbf{N}\}$ is dense in X.

---

[2] This omits $x(0)$ entirely, but it is easy to see that any individual element of the sequence has no impact on the asymptotic equidistribution.

1.1. Equidistribution in tori

Exercise 1.1.3. Let $x\colon \mathbf{N} \to X$ be a sequence into a compact metric space X which is equidistributed relative to some probability measure μ. Let $\varphi\colon \mathbf{R} \to \mathbf{R}$ be a compactly supported, piecewise continuous function with only finitely many pieces. Show that for any $f \in C(X)$ one has

$$\lim_{N \to \infty} \frac{1}{N} \sum_{n \in \mathbf{N}} \varphi(n/N) f(x(n)) = \left(\int_X f \, d\mu \right) \left(\int_0^\infty \varphi(t) \, dt \right)$$

and for any open U whose boundary has measure zero, one has

$$\lim_{N \to \infty} \frac{1}{N} \sum_{n \in \mathbf{N}: x(n) \in U} \varphi(n/N) = \mu(U) \left(\int_0^\infty \varphi(t) \, dt \right).$$

In this section, X will be a torus (i.e., a compact connected abelian Lie group), which from the theory of Lie groups is isomorphic to the standard torus $\mathbf{T}^d$, where d is the dimension of the torus. This torus is then equipped with *Haar measure*, which is the unique Borel probability measure on the torus which is translation-invariant. One can identify the standard torus $\mathbf{T}^d$ with the standard fundamental domain $[0,1)^d$, in which case the Haar measure is equated with the usual Lebesgue measure. We shall call a sequence $x_1, x_2, \ldots$ in $\mathbf{T}^d$ (asymptotically) *equidistributed* if it is (asymptotically) equidistributed with respect to Haar measure.

We have a simple criterion for when a sequence is asymptotically equidistributed, that reduces the problem to that of estimating exponential sums:

Proposition 1.1.2 (Weyl equidistribution criterion). *Let $x\colon \mathbf{N} \to \mathbf{T}^d$. Then x is asymptotically equidistributed if and only if*

$$\lim_{N \to \infty} \mathbf{E}_{n \in [N]} e(k \cdot x(n)) = 0 \tag{1.2}$$

for all $k \in \mathbf{Z}^d \setminus \{0\}$, where $e(y) := e^{2\pi i y}$. Here we use the dot product

$$(k_1, \ldots, k_d) \cdot (x_1, \ldots, x_d) := k_1 x_1 + \cdots + k_d x_d$$

which maps $\mathbf{Z}^d \times \mathbf{T}^d$ to $\mathbf{T}$.

Proof. The "only if" part is immediate from (1.1). For the "if" part, we see from (1.2) that (1.1) holds whenever f is a plane wave $f(y) := e(k \cdot y)$ for some $k \in \mathbf{Z}^d$ (checking the $k = 0$ case separately), and thus by linearity whenever f is a trigonometric polynomial. But by Fourier analysis (or from the *Stone-Weierstrass theorem*), the trigonometric polynomials are dense in $C(\mathbf{T}^d)$ in the uniform topology. The claim now follows from a standard limiting argument. $\square$

As one consequence of this proposition, one can reduce multidimensional equidistribution to single-dimensional equidistribution:

Corollary 1.1.3. *Let $x\colon \mathbf{N} \to \mathbf{T}^d$. Then x is asymptotically equidistributed in $\mathbf{T}^d$ if and only if, for each $k \in \mathbf{Z}^d\setminus\{0\}$, the sequence $n \mapsto k \cdot x(n)$ is asymptotically equidistributed in $\mathbf{T}$.*

Exercise 1.1.4. Show that a sequence $x : \mathbf{N} \to \mathbf{T}^d$ is totally asymptotically equidistributed if and only if one has

$$\lim_{N \to \infty} \mathbf{E}_{n \in [N]} e(k \cdot x(n)) e(\alpha n) = 0 \tag{1.3}$$

for all $k \in \mathbf{Z}^d\setminus\{0\}$ and all rational α.

This quickly gives a test for equidistribution for linear sequences, sometimes known as the *equidistribution theorem*:

Exercise 1.1.5. Let $\alpha, \beta \in \mathbf{T}^d$. By using the geometric series formula, show that the following are equivalent:

 (i) The sequence $n \mapsto n\alpha + \beta$ is asymptotically equidistributed on $\mathbf{N}$.
 (ii) The sequence $n \mapsto n\alpha + \beta$ is totally asymptotically equidistributed on $\mathbf{N}$.
 (iii) The sequence $n \mapsto n\alpha + \beta$ is totally asymptotically equidistributed on $\mathbf{Z}$.
 (iv) α is *irrational*, in the sense that $k \cdot \alpha \neq 0$ for any non-zero $k \in \mathbf{Z}^d$.

Remark 1.1.4. One can view Exercise 1.1.5 as an assertion that a linear sequence x_n will equidistribute itself unless there is an "obvious" algebraic obstruction to it doing so, such as $k \cdot x_n$ being constant for some non-zero k. This theme of algebraic obstructions being the "only" obstructions to uniform distribution will be present throughout the text.

Exercise 1.1.5 shows that linear sequences with irrational shift α are equidistributed. At the other extreme, if α is *rational* in the sense that $m\alpha = 0$ for some positive integer m, then the sequence $n \mapsto n\alpha + \beta$ is clearly periodic of period m, and definitely not equidistributed.

In the one-dimensional case $d = 1$, these are the only two possibilities. But in higher dimensions, one can have a mixture of the two extremes, that exhibits irrational behaviour in some directions and periodic behaviour in others. Consider for instance the two-dimensional sequence $n \mapsto (\sqrt{2}n, \frac{1}{2}n) \bmod \mathbf{Z}^2$. The first coordinate is totally asymptotically equidistributed in $\mathbf{T}$, while the second coordinate is periodic; the shift $(\sqrt{2}, \frac{1}{2})$ is neither irrational nor rational, but is a mixture of both. As such, we see that the two-dimensional sequence is equidistributed with respect to Haar measure on the group $\mathbf{T} \times (\frac{1}{2}\mathbf{Z}/\mathbf{Z})$.

This phenomenon generalises:

1.1. Equidistribution in tori

Proposition 1.1.5 (Equidistribution for abelian linear sequences). *Let T be a torus, and let $x(n) := n\alpha + \beta$ for some $\alpha, \beta \in T$. Then there exists a decomposition $x = x' + x''$, where $x'(n) := n\alpha'$ is totally asymptotically equidistributed on $\mathbf{Z}$ in a subtorus T' of T (with $\alpha' \in T'$, of course), and $x''(n) = n\alpha'' + \beta$ is periodic (or equivalently, that $\alpha'' \in T$ is rational).*

Proof. We induct on the dimension d of the torus T. The claim is vacuous for $d = 0$, so suppose that $d \geq 1$ and that the claim has already been proven for tori of smaller dimension. Without loss of generality we may identify T with $\mathbf{T}^d$.

If α is irrational, then we are done by Exercise 1.1.5, so we may assume that α is not irrational; thus $k \cdot \alpha = 0$ for some non-zero $k \in \mathbf{Z}^d$. We then write $k = mk'$, where m is a positive integer and $k' \in \mathbf{Z}^d$ is *irreducible* (i.e., k' is not a proper multiple of any other element of $\mathbf{Z}^d$); thus $k' \cdot \alpha$ is rational. We may thus write $\alpha = \alpha_1 + \alpha_2$, where α_2 is rational, and $k' \cdot \alpha_1 = 0$. Thus, we can split $x = x_1 + x_2$, where $x_1(n) := n\alpha_1$ and $x_2(n) := n\alpha_2 + \beta$. Clearly x_2 is periodic, while x_1 takes values in the subtorus $T_1 := \{y \in T : k' \cdot y = 0\}$ of T. The claim now follows by applying the induction hypothesis to T_1 (and noting that the sum of two periodic sequences is again periodic). $\square$

As a corollary of the above proposition, we see that any linear sequence $n \mapsto n\alpha + \beta$ in a torus T is equidistributed in some union of finite cosets of a subtorus T'. It is easy to see that this torus T is uniquely determined by α, although there is a slight ambiguity in the decomposition $x = x' + x''$ because one can add or subtract a periodic linear sequence taking values in T from x' and add it to x'' (or vice versa).

Having discussed the linear case, we now consider the more general situation of *polynomial* sequences in tori. To get from the linear case to the polynomial case, the fundamental tool is

Lemma 1.1.6 (van der Corput inequality). *Let $a_1, a_2, \ldots$ be a sequence of complex numbers of magnitude at most 1. Then for every $1 \leq H \leq N$, we have*

$$|\mathbf{E}_{n \in [N]} a_n| \ll \left(\mathbf{E}_{h \in [H]} |\mathbf{E}_{n \in [N]} a_{n+h} \overline{a_n}|\right)^{1/2} + \frac{1}{H^{1/2}} + \frac{H^{1/2}}{N^{1/2}}.$$

Proof. For each $h \in [H]$, we have

$$\mathbf{E}_{n \in [N]} a_n = \mathbf{E}_{n \in [N]} a_{n+h} + O\left(\frac{H}{N}\right)$$

and hence, on averaging,

$$\mathbf{E}_{n \in [N]} a_n = \mathbf{E}_{n \in [N]} \mathbf{E}_{h \in [H]} a_{n+h} + O\left(\frac{H}{N}\right).$$

Applying Cauchy-Schwarz, we conclude

$$\mathbf{E}_{n\in[N]}a_n \ll (\mathbf{E}_{n\in[N]}|\mathbf{E}_{h\in[H]}a_{n+h}|^2)^{1/2} + \frac{H}{N}.$$

We expand out the left-hand side as

$$\mathbf{E}_{n\in[N]}a_n \ll (\mathbf{E}_{h,h'\in[H]}\mathbf{E}_{n\in[N]}a_{n+h}\overline{a_{n+h'}})^{1/2} + \frac{H}{N}.$$

The diagonal contribution $h = h'$ is $O(1/H)$. By symmetry, the off-diagonal contribution can be dominated by the contribution when $h > h'$. Making the change of variables $n \mapsto n - h'$, $h \mapsto h + h'$ (accepting a further error of $O(H^{1/2}/N^{1/2})$), we obtain the claim. □

Corollary 1.1.7 (van der Corput lemma). *Let $x \colon \mathbf{N} \to \mathbf{T}^d$ be such that the derivative sequence $\partial_h x \colon n \mapsto x(n+h) - x(n)$ is asymptotically equidistributed on $\mathbf{N}$ for all positive integers h. Then x_n is asymptotically equidistributed on $\mathbf{N}$. Similarly with $\mathbf{N}$ replaced by $\mathbf{Z}$.*

Proof. We just prove the claim for $\mathbf{N}$, as the claim for $\mathbf{Z}$ is analogous (and can in any case be deduced from the $\mathbf{N}$ case).

By Proposition 1.1.2, we need to show that for each non-zero $k \in \mathbf{Z}^d$, the exponential sum

$$|\mathbf{E}_{n\in[N]}e(k \cdot x(n))|$$

goes to zero as $N \to \infty$. Fix an $H > 0$. By Lemma 1.1.6, this expression is bounded by

$$\ll (\mathbf{E}_{h\in[H]}|\mathbf{E}_{n\in[N]}e(k \cdot (x(n+h) - x(n)))|)^{1/2} + \frac{1}{H^{1/2}} + \frac{H^{1/2}}{N^{1/2}}.$$

On the other hand, for each fixed positive integer h, we have from hypothesis and Proposition 1.1.2 that $|\mathbf{E}_{n\in[N]}e(k \cdot (x(n+h) - x(n)))|$ goes to zero as $N \to \infty$. Taking limit superior as $N \to \infty$, we conclude that

$$\limsup_{N\to\infty} |\mathbf{E}_{n\in[N]}e(k \cdot x(n))| \ll \frac{1}{H^{1/2}}.$$

Since H is arbitrary, the claim follows. □

Remark 1.1.8. There is another famous lemma by van der Corput concerning oscillatory integrals, but it is not directly related to the material discussed here.

Corollary 1.1.7 has the following immediate corollary:

Corollary 1.1.9 (Weyl equidistribution theorem for polynomials). *Let $s \geq 1$ be an integer, and let $P(n) = \alpha_s n^s + \cdots + \alpha_0$ be a polynomial of degree s with $\alpha_0, \ldots, \alpha_s \in \mathbf{T}^d$. If α_s is irrational, then $n \mapsto P(n)$ is asymptotically equidistributed on $\mathbf{Z}$.*

1.1. Equidistribution in tori

Proof. We induct on s. For $s = 1$ this follows from Exercise 1.1.5. Now suppose that $s > 1$, and that the claim has already been proven for smaller values of s. For any positive integer h, we observe that $P(n+h) - P(n)$ is a polynomial of degree $s - 1$ in n with leading coefficient $sh\alpha_s n^{s-1}$. As α_s is irrational, $sh\alpha_s$ is irrational also, and so by the induction hypothesis, $P(n+h) - P(n)$ is asymptotically equidistributed. The claim now follows from Corollary 1.1.7. $\square$

Exercise 1.1.6. Let $P(n) = \alpha_s n^s + \cdots + \alpha_0$ be a polynomial of degree s in $\mathbf{T}^d$. Show that the following are equivalent:

(i) P is asymptotically equidistributed on $\mathbf{N}$.

(ii) P is totally asymptotically equidistributed on $\mathbf{N}$.

(iii) P is totally asymptotically equidistributed on $\mathbf{Z}$.

(iv) There does not exist a non-zero $k \in \mathbf{Z}^d$ such that $k \cdot \alpha_1 = \cdots = k \cdot \alpha_s = 0$.

(*Hint:* It is convenient to first use Corollary 1.1.3 to reduce to the one-dimensional case.)

This gives a polynomial variant of the equidistribution theorem:

Exercise 1.1.7 (Equidistribution theorem for abelian polynomial sequences). Let T be a torus, and let P be a polynomial map from $\mathbf{Z}$ to T of some degree $s \geq 0$. Show that there exists a decomposition $P = P' + P''$, where P', P'' are polynomials of degree s, P' is totally asymptotically equidistributed in a subtorus T' of T on $\mathbf{Z}$, and P'' is periodic (or equivalently, that all non-constant coefficients of P'' are rational).

In particular, we see that polynomial sequences in a torus are equidistributed with respect to a finite combination of Haar measures of cosets of a subtorus. Note that this finite combination can have multiplicity; for instance, when considering the polynomial map $n \mapsto (\sqrt{2}n, \frac{1}{3}n^2) \bmod \mathbf{Z}^2$, it is not hard to see that this map is equidistributed with respect to $1/3$ times the Haar probability measure on $(\mathbf{T}) \times \{0 \bmod \mathbf{Z}\}$, plus $2/3$ times the Haar probability measure on $(\mathbf{T}) \times \{\frac{1}{3} \bmod \mathbf{Z}\}$.

Exercise 1.1.7 gives a satisfactory description of the asymptotic equidistribution of arbitrary polynomial sequences in tori. We give just one example of how such a description can be useful:

Exercise 1.1.8 (Recurrence). Let T be a torus, let P be a polynomial map from $\mathbf{Z}$ to T, and let n_0 be an integer. Show that there exists a sequence n_j of positive integers going to infinity such that $P(n_j) \to P(n_0)$.

We discussed recurrence for one-dimensional sequences $x\colon n \mapsto x(n)$. It is also of interest to establish an analogous theory for multi-dimensional sequences, as follows.

Definition 1.1.10. A multidimensional sequence $x\colon \mathbf{Z}^m \to X$ is *asymptotically equidistributed* relative to a probability measure μ if, for every continuous, compactly supported function $\varphi\colon \mathbf{R}^m \to \mathbf{R}$ and every function $f \in C(X)$, one has

$$\frac{1}{N^m} \sum_{n \in \mathbf{Z}^m} \varphi(n/N) f(x(n)) \to \left(\int_{\mathbf{R}^m} \varphi\right)\left(\int_X f \, d\mu\right)$$

as $N \to \infty$. The sequence is *totally asymptotically equidistributed* relative to μ if the sequence $n \mapsto x(qn+r)$ is asymptotically equidistributed relative to μ for all positive integers q and all $r \in \mathbf{Z}^m$.

Exercise 1.1.9. Show that this definition of equidistribution on $\mathbf{Z}^m$ coincides with the preceding definition of equidistribution on $\mathbf{Z}$ in the one-dimensional case $m = 1$.

Exercise 1.1.10 (Multidimensional Weyl equidistribution criterion)**.** Let $x\colon \mathbf{Z}^m \to \mathbf{T}^d$ be a multidimensional sequence. Show that x is asymptotically equidistributed if and only if

$$(1.4) \qquad \lim_{N \to \infty} \frac{1}{N^m} \sum_{n \in \mathbf{Z}^m : n/N \in B} e(k \cdot x(n)) = 0$$

for all $k \in \mathbf{Z}^d \setminus \{0\}$ and all rectangular boxes B in $\mathbf{R}^m$. Then show that x is totally asymptotically equidistributed if and only if

$$(1.5) \qquad \lim_{N \to \infty} \frac{1}{N^m} \sum_{n \in \mathbf{Z}^m : n/N \in B} e(k \cdot x(n)) e(\alpha \cdot n) = 0$$

for all $k \in \mathbf{Z}^d \setminus \{0\}$, all rectangular boxes B in $\mathbf{R}^m$, and all rational $\alpha \in \mathbf{Q}^m$.

Exercise 1.1.11. Let $\alpha_1, \ldots, \alpha_m, \beta \in \mathbf{T}^d$, and let $x\colon \mathbf{Z}^m \to \mathbf{T}^d$ be the linear sequence $x(n_1, \ldots, n_m) := n_1 \alpha_1 + \cdots + n_m \alpha_m + \beta$. Show that the following are equivalent:

  (i) The sequence x is asymptotically equidistributed on $\mathbf{Z}^m$.
  (ii) The sequence x is totally asymptotically equidistributed on $\mathbf{Z}^m$.
  (iii) We have $(k \cdot \alpha_1, \ldots, k \cdot \alpha_m) \neq 0$ for any non-zero $k \in \mathbf{Z}^d$.

Exercise 1.1.12 (Multidimensional van der Corput lemma)**.** Let $x\colon \mathbf{Z}^m \to \mathbf{T}^d$ be such that the sequence $\partial_h x\colon n \mapsto x(n+h) - x(n)$ is asymptotically equidistributed on $\mathbf{Z}^m$ for all h outside of a hyperplane in $\mathbf{R}^m$. Show that x is asymptotically equidistributed on $\mathbf{Z}^m$.

1.1. Equidistribution in tori

Exercise 1.1.13. Let
$$P(n_1,\ldots,n_m) := \sum_{i_1,\ldots,i_m \geq 0 : i_1+\cdots+i_m \leq s} \alpha_{i_1,\ldots,i_m} n_1^{i_1} \ldots n_m^{i_m}$$
be a polynomial map from $\mathbf{Z}^m$ to $\mathbf{T}^d$ of degree s, where $\alpha_{i_1,\ldots,i_m} \in \mathbf{T}^d$ are coefficients. Show that the following are equivalent:

  (i) P is asymptotically equidistributed on $\mathbf{Z}^m$.
  (ii) P is totally asymptotically equidistributed on $\mathbf{Z}^m$.
  (iii) There does not exist a non-zero $k \in \mathbf{Z}^d$ such that $k \cdot \alpha_{i_1,\ldots,i_m} = 0$ for all $(i_1,\ldots,i_m) \neq 0$.

Exercise 1.1.14 (Equidistribution for abelian multidimensional polynomial sequences). Let T be a torus, and let P be a polynomial map from $\mathbf{Z}^m$ to T of some degree $s \geq 0$. Show that there exists a decomposition $P = P' + P''$, where P', P'' are polynomials of degree s, P' is totally asymptotically equidistributed in a subtorus T' of T on $\mathbf{Z}^m$, and P'' is periodic with respect to some finite index sublattice of $\mathbf{Z}^m$ (or equivalently, that all non-constant coefficients of P'' are rational).

We give just one application of this multidimensional theory, that gives a hint as to why the theory of equidistribution of polynomials may be relevant:

Exercise 1.1.15. Let T be a torus, let P be a polynomial map from $\mathbf{Z}$ to T, let $\varepsilon > 0$, and let $k \geq 1$. Show that there exists positive integers $a, r \geq 1$ such that $P(a), P(a+r), \ldots, P(a+(k-1)r)$ all lie within ε of each other. (*Hint:* Consider the polynomial map from $\mathbf{Z}^2$ to T^k that maps (a,r) to $(P(a),\ldots,P(a+(k-1)r))$. One can also use the one-dimensional theory by freezing a and only looking at the equidistribution in r.)

1.1.2. Single-scale equidistribution theory. We now turn from the asymptotic equidistribution theory to the equidistribution theory at a single scale N. Thus, instead of analysing the qualitative distribution of infinite sequence $x \colon \mathbf{N} \to X$, we consider instead the quantitative distribution of a finite sequence $x \colon [N] \to X$, where N is a (large) natural number and $[N] := \{1,\ldots,N\}$. To make everything quantitative, we will replace the notion of a continuous function by that of a *Lipschitz function*. Recall that the (inhomogeneous) Lipschitz norm $\|f\|_{\mathrm{Lip}}$ of a function $f \colon X \to \mathbf{R}$ on a metric space $X = (X,d)$ is defined by the formula
$$\|f\|_{\mathrm{Lip}} := \sup_{x \in X} |f(x)| + \sup_{x,y \in X : x \neq y} \frac{|f(x)-f(y)|}{d(x,y)}.$$
We also define the homogeneous Lipschitz semi-norm
$$\|f\|_{\dot{\mathrm{Lip}}} := \sup_{x,y \in X : x \neq y} \frac{|f(x)-f(y)|}{d(x,y)}.$$

Definition 1.1.11. Let $X = (X, d)$ be a compact metric space, let $\delta > 0$, let μ be a probability measure on X. A finite sequence $x \colon [N] \to X$ is said to be *δ-equidistributed* relative to μ if one has

$$(1.6) \qquad |\mathbf{E}_{n \in [N]} f(x(n)) - \int_X f \, d\mu| \leq \delta \|f\|_{\mathrm{Lip}}$$

for all Lipschitz functions $f \colon X \to \mathbf{R}$.

We say that the sequence $x_1, \ldots, x_N \in X$ is *totally δ-equidistributed* relative to μ if one has

$$|\mathbf{E}_{n \in P} f(x(n)) - \int_X f \, d\mu| \leq \delta \|f\|_{\mathrm{Lip}}$$

for all Lipschitz functions $f \colon X \to \mathbf{R}$ and all arithmetic progressions P in $[N]$ of length at least δN.

In this section, we will only apply this concept to the torus $\mathbf{T}^d$ with the Haar measure μ and the metric inherited from the Euclidean metric. However, in subsequent sections we will also consider equidistribution in other spaces, most notably on *nilmanifolds*.

Exercise 1.1.16. Let $x(1), x(2), x(3), \ldots$ be a sequence in a metric space $X = (X, d)$, and let μ be a probability measure on X. Show that the sequence $x(1), x(2), \ldots$ is asymptotically equidistributed relative to μ if and only if, for every $\delta > 0$, $x(1), \ldots, x(N)$ is δ-equidistributed relative to μ whenever N is sufficiently large depending on δ, or equivalently if $x(1), \ldots, x(N)$ is $\delta(N)$-equidistributed relative to μ for all $N > 0$, where $\delta(N) \to 0$ as $N \to \infty$. (*Hint:* You will need the *Arzelá-Ascoli theorem*.)

Similarly, show that $x(1), x(2), \ldots$ is totally asymptotically equidistributed relative to μ if and only if, for every $\delta > 0$, $x(1), \ldots, x(N)$ is totally δ-equidistributed relative to μ whenever N is sufficiently large depending on δ, or equivalently if $x(1), \ldots, x(N)$ is totally $\delta(N)$-equidistributed relative to μ for all $N > 0$, where $\delta(N) \to 0$ as $N \to \infty$.

Remark 1.1.12. More succinctly, (total) asymptotic equidistribution of $x(1), x(2), \ldots$ is equivalent to (total) $o_{N \to \infty}(1)$-equidistribution of $x(1), \ldots, x(N)$ as $N \to \infty$, where $o_{n \to \infty}(1)$ denotes a quantity that goes to zero as $N \to \infty$. Thus we see that asymptotic notation such as $o_{n \to \infty}(1)$ can efficiently conceal a surprisingly large number of quantifiers.

Exercise 1.1.17. Let N_0 be a large integer, and let $x(n) := n/N_0 \bmod 1$ be a sequence in the standard torus $\mathbf{T} = \mathbf{R}/\mathbf{Z}$ with Haar measure. Show that whenever N is a positive multiple of N_0, then the sequence $x(1), \ldots, x(N)$ is $O(1/N_0)$-equidistributed. What happens if N is not a multiple of N_0?

If, furthermore, $N \geq N_0^2$, show that $x(1), \ldots, x(N)$ is $O(1/\sqrt{N_0})$-equidistributed. Why is a condition such as $N \geq N_0^2$ necessary?

1.1. Equidistribution in tori

Note that the above exercise does not specify the exact relationship between δ and N when one is given an asymptotically equidistributed sequence $x(1), x(2), \ldots$; this relationship is the additional piece of information provided by single-scale equidistribution that is not present in asymptotic equidistribution.

It turns out that much of the asymptotic equidistribution theory has a counterpart for single-scale equidistribution. We begin with the Weyl criterion.

Proposition 1.1.13 (Single-scale Weyl equidistribution criterion). *Let x_1, $x_2, \ldots, x_N$ be a sequence in $\mathbf{T}^d$, and let $0 < \delta < 1$.*

(i) *If $x_1, \ldots, x_N$ is δ-equidistributed, and $k \in \mathbf{Z}^d \setminus \{0\}$ has magnitude $|k| \leq \delta^{-c}$, then one has*
$$|\mathbf{E}_{n \in [N]} e(k \cdot x_n)| \ll_d \delta^c$$
if $c > 0$ is a small enough absolute constant.

(ii) *Conversely, if $x_1, \ldots, x_N$ is not δ-equidistributed, then there exists $k \in \mathbf{Z}^d \setminus \{0\}$ with magnitude $|k| \ll_d \delta^{-C_d}$, such that*
$$|\mathbf{E}_{n \in [N]} e(k \cdot x_n)| \gg_d \delta^{C_d}$$
for some C_d depending on d.

Proof. The first claim is immediate as the function $x \mapsto e(k \cdot x)$ has mean zero and Lipschitz constant $O_d(|k|)$, so we turn to the second claim. By hypothesis, (1.6) fails for some Lipschitz f. We may subtract off the mean and assume that $\int_{\mathbf{T}^d} f = 0$; we can then normalise the Lipschitz norm to be one; thus we now have
$$|\mathbf{E}_{n \in [N]} f(x_n)| > \delta.$$
We introduce a summation parameter $R \in \mathbf{N}$, and consider the *Fejér partial Fourier series*
$$F_R f(x) := \sum_{k \in \mathbf{Z}^d} m_R(k) \hat{f}(k) e(k \cdot x)$$
where $\hat{f}(k)$ are the Fourier coefficients
$$\hat{f}(k) := \int_{\mathbf{T}^d} f(x) e(-k \cdot x) \, dx$$
and m_R is the Fourier multiplier
$$m_R(k_1, \ldots, k_d) := \prod_{j=1}^{d} \left(1 - \frac{|k_j|}{R}\right)_+.$$

Standard Fourier analysis shows that we have the convolution representation
$$F_R f(x) = \int_{\mathbf{T}^d} f(y) K_R(x-y)$$
where K_R is the Fejér kernel
$$K_R(x_1, \ldots, x_d) := \prod_{j=1}^{d} \frac{1}{R} \left(\frac{\sin(\pi R x_j)}{\sin(\pi x_j)} \right)^2.$$
Using the kernel bounds
$$\int_{\mathbf{T}^d} K_R = 1$$
and
$$|K_R(x)| \ll_d \prod_{j=1}^{d} R(1 + R\|x_j\|_{\mathbf{T}})^{-2},$$
where $\|x\|_{\mathbf{T}}$ is the distance from x to the nearest integer, and the Lipschitz nature of f, we see that
$$F_R f(x) = f(x) + O_d(1/R).$$
Thus, if we choose R to be a sufficiently small multiple of $1/\delta$ (depending on d), one has
$$|\mathbf{E}_{n \in [N]} F_R f(x_n)| \gg \delta$$
and thus by the pigeonhole principle (and the trivial bound $\hat{f}(k) = O(1)$ and $\hat{f}(0) = 0$) we have
$$|\mathbf{E}_{n \in [N]} e(k \cdot x_n)| \gg_d \delta^{O_d(1)}$$
for some non-zero k of magnitude $|k| \ll_d \delta^{-O_d(1)}$, and the claim follows. □

There is an analogue for total equidistribution:

Exercise 1.1.18. Let $x_1, x_2, \ldots, x_N$ be a sequence in $\mathbf{T}^d$, and let $0 < \delta < 1$.

(i) If $x_1, \ldots, x_N$ is totally δ-equidistributed, $k \in \mathbf{Z}^d \backslash \{0\}$ has magnitude $|k| \leq \delta^{-c_d}$, and a is a rational of height at most δ^{-c_d}, then one has
$$|\mathbf{E}_{n \in [N]} e(k \cdot x_n) e(an)| \ll_d \delta^{c_d}$$
if $c_d > 0$ is a small enough constant depending only on d.

(ii) Conversely, if $x_1, \ldots, x_N$ is *not* totally δ-equidistributed, then there exists $k \in \mathbf{Z}^d \backslash \{0\}$ with magnitude $|k| \ll_d \delta^{-C_d}$, and a rational a of height $O_d(\delta^{-C_d})$, such that
$$|\mathbf{E}_{n \in [N]} e(k \cdot x_n) e(an)| \gg_d \delta^{C_d}$$
for some C_d depending on d.

This gives a version of Exercise 1.1.5:

Exercise 1.1.19. Let $\alpha, \beta \in \mathbf{T}^d$, let $N \geq 1$, and let $0 < \delta < 1$. Suppose that the linear sequence $(\alpha n + \beta)_{n=1}^N$ is not totally δ-equidistributed. Show that there exists a non-zero $k \in \mathbf{Z}^d$ with $|k| \ll_d \delta^{-O_d(1)}$ such that $\|k \cdot \alpha\|_\mathbf{T} \ll_d \delta^{-O_d(1)}/N$.

Next, we give an analogue of Corollary 1.1.7:

Exercise 1.1.20 (Single-scale van der Corput lemma). Let $x_1, x_2, \ldots, x_N \in \mathbf{T}^d$ be a sequence which is not totally δ-equidistributed for some $0 < \delta \leq 1/2$. Let $1 \leq H \leq \delta^{-C_d} N$ for some sufficiently large C_d depending only on d. Then there exists at least $\delta^{C_d} H$ integers $h \in [-H, H]$ such that the sequence $(x_{n+h} - x_n)_{n=1}^N$ is not totally δ^{C_d}-equidistributed (where we extend x_n by zero outside of $\{1, \ldots, N\}$). (*Hint:* Apply Lemma 1.1.6.)

Just as in the asymptotic setting, we can use the van der Corput lemma to extend the linear equidistribution theory to polynomial sequences. To get satisfactory results, though, we will need an additional input, namely the following classical lemma, essentially due to Vinogradov:

Lemma 1.1.14. *Let* $\alpha \in \mathbf{T}$, $0 < \varepsilon < 1/100$, $100\varepsilon < \delta < 1$, *and* $N \geq 100/\delta$. *Suppose that* $\|n\alpha\|_\mathbf{T} \leq \varepsilon$ *for at least* δN *values of* $n \in [-N, N]$. *Then there exists a positive integer* $q = O(1/\delta)$ *such that* $\|\alpha q\|_\mathbf{T} \ll \frac{\varepsilon q}{\delta N}$.

The key point here is that one starts with many multiples of α being somewhat close ($O(\varepsilon)$) to an integer, but concludes that there is a single multiple of α which is *very* close ($O(\varepsilon/N)$, ignoring factors of δ) to an integer.

Proof. By the pigeonhole principle, we can find two distinct integers $n, n' \in [-N, N]$ with $|n - n'| \ll 1/\delta$ such that $\|n\alpha\|_\mathbf{T}, \|n'\alpha\|_\mathbf{T} \leq \varepsilon$. Setting $q := |n' - n|$, we thus have $\|q\alpha\|_\mathbf{T} \leq 2\varepsilon$. We may assume that $q\alpha \neq 0$ since we are done otherwise. Since $N \geq 100/\delta$, we have $N/q \geq 10$ (say).

Now partition $[-N, N]$ into q arithmetic progressions $\{nq + r : -N/q + O(1) \leq n \leq N/q + O(1)\}$ for some $r = 0, \ldots, q-1$. By the pigeonhole principle, there must exist an r for which the set

$$\{-N/q + O(1) \leq n \leq N/q + O(1) : \|\alpha(nq + r)\|_\mathbf{T} \leq \varepsilon\}$$

has cardinality at least $\delta N/q$. On the other hand, since $\|q\alpha\|_\mathbf{T} \leq 2\varepsilon \leq 0.02$, we see that this set consists of intervals of length at most $2\varepsilon/\|q\alpha\|_\mathbf{T}$, punctuated by gaps of length at least $0.9/\|q\alpha\|_\mathbf{T}$ (say). Since the gaps are at least $0.45/\varepsilon$ times as large as the intervals, we see that if two or more of these intervals appear in the set, then the cardinality of the set is at most $100\varepsilon N/q < \delta N/q$, a contradiction. Thus at most one interval appears in the set, which implies that $2\varepsilon/\|q\alpha\|_\mathbf{T} \geq \delta N/q$, and the claim follows. $\square$

Remark 1.1.15. The numerical constants can of course be improved, but this is not our focus here.

Exercise 1.1.21. Let $P\colon \mathbf{Z} \to \mathbf{T}^d$ be a polynomial sequence $P(n) := \alpha_s n^s + \cdots + \alpha_0$, let $N \geq 1$, and let $0 < \delta < 1$. Suppose that the polynomial sequence P is not totally δ-equidistributed on $[N]$. Show that there exists a non-zero $k \in \mathbf{Z}^d$ with $|k| \ll_{d,s} \delta^{-O_{d,s}(1)}$ such that $\|k \cdot \alpha_s\|_{\mathbf{T}} \ll_{d,s} \delta^{-O_{d,s}(1)}/N^s$. (*Hint:* Induct on s starting with Exercise 1.1.19 for the base case, and then using Exercise 1.1.20 and Lemma 1.1.14 to continue the induction.)

Note the N^s denominator; the higher-degree coefficients of a polynomial need to be *very* rational in order not to cause equidistribution.

The above exercise only controls the top degree coefficient, but we can in fact control all coefficients this way:

Lemma 1.1.16. *With the hypotheses of Exercise 1.1.21, we can in fact find a non-zero $k \in \mathbf{Z}^d$ with $|k| \ll_{d,s} \delta^{-O_{d,s}(1)}$ such that $\|k \cdot \alpha_i\|_{\mathbf{T}} \ll_{d,s} \delta^{-O_{d,s}(1)}/N^i$ for all $i = 0, \ldots, s$.*

Proof. We shall just establish the one-dimensional case $d = 1$, as the general dimensional case then follows from Exercise 1.1.18.

The case $s \leq 1$ follows from Exercise 1.1.19, so assume inductively that $s > 1$ and that the claim has already been proven for smaller values of s. We allow all implied constants to depend on s. From Exercise 1.1.21, we already can find a positive k with $k = O(\delta^{-O(1)})$ such that $\|k\alpha_s\|_{\mathbf{T}} \ll \delta^{-O(1)}/N^s$. We now partition $[N]$ into arithmetic progressions of spacing k and length $N' \sim \delta^C N$ for some sufficiently large C; then by the pigeonhole principle, we see that P fails to be totally $\gg \delta^{O(1)}$-equidistributed on one of these progressions. But on one such progression (which can be identified with $[N']$) the degree s component of P is essentially constant (up to errors much smaller than δ) if C is large enough; if one then applies the induction hypothesis to the remaining portion of P on this progression, we can obtain the claim. $\square$

This gives us the following analogue of Exercise 1.1.7. We say that a subtorus T of some dimension d' of a standard torus $\mathbf{T}^d$ has *complexity* at most M if there exists an invertible linear transformation $L \in SL_d(\mathbf{Z})$ with integer coefficients (which can thus be viewed as a homeomorphism of $\mathbf{T}^d$ that maps T to the standard torus $\mathbf{T}^{d'} \times \{0\}^{d-d'}$), and such that all coefficients have magnitude at most M.

Exercise 1.1.22. Show that every subtorus (i.e., compact connected Lie subgroup) T of $\mathbf{T}^d$ has finite complexity. (*Hint:* Let V be the Lie algebra of T, then identify V with a subspace of $\mathbf{R}^d$ and T with $V/(V \cap \mathbf{Z}^d)$. Show

1.1. Equidistribution in tori

that $V \cap \mathbf{Z}^d$ is a full rank sublattice of V, and is thus generated by $\dim(V)$ independent generators.)

Proposition 1.1.17 (Single-scale equidistribution theorem for abelian polynomial sequences)**.** *Let P be a polynomial map from $\mathbf{Z}$ to $\mathbf{T}^d$ of some degree $s \geq 0$, and let $F \colon \mathbf{R}^+ \to \mathbf{R}^+$ be an increasing function. Then there exists an integer $1 \leq M \leq O_{F,s,d}(1)$ and a decomposition*

$$P = P_{\mathrm{smth}} + P_{\mathrm{equi}} + P_{\mathrm{rat}}$$

into polynomials of degree s, where

  (i) *(P_{smth} is smooth) The ith coefficient $\alpha_{i,\mathrm{smth}}$ of P_{smth} has size $O(M/N^i)$. In particular, on the interval $[N]$, P_{smth} is Lipschitz with homogeneous norm $O_{s,d}(M/N)$.*

  (ii) *(P_{equi} is equidistributed) There exists a subtorus T of $\mathbf{T}^d$ of complexity at most M and some dimension d', such that P_{equi} takes values in T and is totally $1/F(M)$-equidistributed on $[N]$ in this torus (after identifying this torus with $\mathbf{T}^{d'}$ using an invertible linear transformation of complexity at most M).*

  (iii) *(P_{rat} is rational) The coefficients $\alpha_{i,\mathrm{rat}}$ of P_{rat} are such that $q\alpha_{i,\mathrm{rat}} = 0$ for some $1 \leq q \leq M$ and all $0 \leq i \leq s$. In particular, $qP_{\mathrm{rat}} = 0$ and P_{rat} is periodic with period q.*

If, furthermore, F is of polynomial growth, and more precisely $F(M) \leq KM^A$ for some $A, K \geq 1$, then one can take $M \ll_{A,s,d} K^{O_{A,s,d}(1)}$.

Example 1.1.18. Consider the linear flow $P(n) := (\sqrt{2}n, (\frac{1}{2}+\frac{1}{N})n) \bmod \mathbf{Z}^2$ in $\mathbf{T}^2$ on $[N]$. This flow can be decomposed into a smooth flow $P_{\mathrm{smth}}(n) := (0, \frac{1}{N}n) \bmod \mathbf{Z}^2$ with a homogeneous Lipschitz norm of $O(1/N)$, an equidistributed flow $P_{\mathrm{equi}}(n) := (\sqrt{2}n, 0) \bmod \mathbf{Z}^2$ which will be δ-equidistributed on the subtorus $\mathbf{T}^1 \times \{0\}$ for a reasonably small δ (in fact one can take δ as small as N^{-c} for some small absolute constant $c > 0$), and a rational flow $P_{\mathrm{rat}}(n) := (0, \frac{1}{2}n) \bmod \mathbf{Z}^2$, which is periodic with period 2. This example illustrates how all three components of this decomposition arise naturally in the single-scale case.

Remark 1.1.19. Comparing this result with the asymptotically equidistributed analogue in Example 1.1.7, we notice several differences. Firstly, we now have the smooth component P_{smth}, which did not previously make an appearance (except implicitly, as the constant term in P'). Secondly, the equidistribution of the component P_{equi} is not infinite, but is the next best thing, namely it is given by an arbitrary function F of the quantity M, which controls the other components of the decomposition.

Proof. The case $s = 0$ is trivial, so suppose inductively that $s \geq 1$, and that the claim has already been proven for lower degrees. Then for fixed degree, the case $d = 0$ is vacuously true, so we make a further inductive assumption $d \geq 1$ and the claim has already been proven for smaller dimensions (keeping s fixed).

If P is already totally $1/F(1)$-equidistributed then we are done (setting $P_{\text{equi}} = P$ and $P_{\text{smth}} = P_{\text{rat}} = 0$ and $M = 1$), so suppose that this is not the case. Applying Exercise 1.1.21, we conclude that there is some non-zero $k \in \mathbf{Z}^d$ with $|k| \ll_{d,s} F(1)^{O_{d,s}(1)}$ such that

$$\|k \cdot \alpha_i\|_{\mathbf{T}} \ll_{d,s} F(1)^{O_{d,s}(1)}/N^i$$

for all $i = 0, \ldots, s$. We split $k = mk'$ where k' is irreducible and m is a positive integer. We can therefore split $\alpha_i = \alpha_{i,\text{smth}} + \alpha_{i,\text{rat}} + \alpha'_i$ where $\alpha_{i,\text{smth}} = O(F(1)^{O_{d,s}(1)}/N^i)$, $q\alpha_i = 0$ for some positive integer $q = O_{d,s}(F(1)^{O_{d,s}(1)})$, and $k' \cdot \alpha'_i = 0$. This then gives a decomposition $P = P_{\text{smth}} + P' + P_{\text{rat}}$, with P' taking values in the subtorus $\{x \in \mathbf{T}^d : k' \cdot x = 0\}$, which can be identified with $\mathbf{T}^{d-1}$ after an invertible linear transformation with integer coefficients of size $O_{d,s}(F(1)^{O_{d,s}(1)})$. If one applies the induction hypothesis to P' (with F replaced by a suitably larger function F') one then obtains the claim.

The final claim about polynomial bounds can be verified by a closer inspection of the argument (noting that all intermediate steps are polynomially quantitative, and that the length of the induction is bounded by $O_{d,s}(1)$). $\square$

Remark 1.1.20. It is instructive to see how this smooth-equidistributed-rational decomposition evolves as N increases. Roughly speaking, the torus T that the P_{equi} component is equidistributed on is stable at most scales, but there will be a finite number of times in which a "growth spurt" occurs and T jumps up in dimension. For instance, consider the linear flow $P(n) := (n/N_0, n/N_0^2) \bmod \mathbf{Z}^2$ on the two-dimensional torus. At scales $N \ll N_0$ (and with F fixed, and N_0 assumed to be sufficiently large depending on F), P consists entirely of the smooth component. But as N increases past N_0, the first component of P no longer qualifies as smooth, and becomes equidistributed instead; thus in the range $N_0 \ll N \ll N_0^2$, we have $P_{\text{smth}}(n) = (0, n/N_0^2) \bmod \mathbf{Z}^2$ and $P_{\text{equi}}(n) = (n/N_0, 0) \bmod \mathbf{Z}^2$ (with P_{rat} remaining trivial), with the torus T increasing from the trivial torus $\{0\}^2$ to $\mathbf{T}^1 \times \{0\}$. A second transition occurs when N exceeds N_0^2, at which point P_{equi} encompasses all of P. Evolving things in a somewhat different direction, if one then increases F so that $F(1)$ is much larger than N_0^2, then P will now entirely consist of a rational component P_{rat}. These sorts of dynamics are not directly seen if one only looks at the asymptotic theory, which roughly speaking is concerned with the limit after taking $N \to \infty$, and *then* taking a second limit by making the growth function F go to infinity.

There is a multidimensional version of Proposition 1.1.17, but we will not describe it here; see [**GrTa2011**] for a statement (and also see the next section for the ultralimit counterpart of this statement).

Remark 1.1.21. These single-scale abelian equidistribution theorems are a special case of a more general single-scale *nilpotent* equidistribution theorem, which will play an important role in later aspects of the theory, and which was the main result of the aforementioned paper of Ben Green and myself.

As an example of this theorem in action, we give a single-scale strengthening of Exercise 1.1.8 (and Exercise 1.1.15):

Exercise 1.1.23 (Recurrence). Let P be a polynomial map from $\mathbf{Z}$ to $\mathbf{T}^d$ of degree s, and let $N \geq 1$ be an integer. Show that for every $\varepsilon > 0$ and $N > 1$, and every integer $n_0 \in [N]$, we have
$$|\{n \in [N] : \|P(n) - P(n_0)\| \leq \varepsilon\}| \gg_{d,s} \varepsilon^{O_{d,s}(1)} N.$$

Exercise 1.1.24 (Multiple recurrence). With the notation of Exercise 1.1.23, establish that
$$|\{r \in [-N, N] : \|P(n_0 + jr) - P(n_0)\| \leq \varepsilon \text{ for } j = 0, 1, \ldots, k-1\}|$$
$$\gg_{d,s,k} \varepsilon^{O_{d,s,k}(1)} N$$
for any $k \geq 1$.

Exercise 1.1.25 (Syndeticity). A set of integers is *syndetic* if it has bounded gaps (or equivalently, if a finite number of translates of this set can cover all of $\mathbf{Z}$). Let $P \colon \mathbf{Z} \to \mathbf{T}^d$ be a polynomial and let $\varepsilon > 0$. Show that the set $\{n \in \mathbf{Z} : \|P(n) - P(n_0)\| \leq \varepsilon\}$ is syndetic. (*Hint:* First reduce to the case when P is (totally) asymptotically equidistributed. Then, if N is large enough, show (by inspection of the proof of Exercise 1.1.21) that the translates $P(\cdot + n_0)$ are ε-equidistributed on $[N]$ uniformly for all $n \in \mathbf{Z}$, for any fixed $\varepsilon > 0$. Note how the asymptotic theory and the single-scale theory need to work together to obtain this result.)

1.1.3. Ultralimit equidistribution theory. The single-scale theory was somewhat more complicated than the asymptotic theory, in part because one had to juggle parameters such as N, δ, and (for the equidistribution theorems) F as well. However, one can clean up this theory somewhat (especially if one does not wish to quantify the dependence of bounds on the equidistribution parameter δ) by using an ultralimit, which causes the δ

and F parameters to disappear, at the cost of converting the finitary theory to an infinitary one. Ultralimit analysis is discussed in Section 2.1; we give a quick review here.

We first fix a *non-principal ultrafilter* $\alpha_\infty \in \beta\mathbf{N}\backslash\mathbf{N}$ (see Section 2.1 for a definition of a non-principal ultrafilter). A property P_α pertaining to a natural number α is said to hold *for all α sufficiently close to α_∞* if the set of α for which P_α holds lies in the ultrafilter α_∞. Two sequences $(x_\alpha)_{\alpha \in \mathbf{N}}, (y_\alpha)_{\alpha \in \mathbf{N}}$ of objects are *equivalent* if one has $x_\alpha = y_\alpha$ for all α sufficiently close to α_∞, and we define the *ultralimit* $\lim_{\alpha \to \alpha_\infty} x_\alpha$ to be the equivalence class of all sequences equivalent to $(x_\alpha)_{\alpha \in \mathbf{N}}$, with the convention that x is identified with its own ultralimit $\lim_{\alpha \to \alpha_\infty} x_\alpha$. Given any sequence X_α of sets, the *ultraproduct* $\prod_{\alpha \to \alpha_\infty} X_\alpha$ is the space of all ultralimits $\lim_{\alpha \to \alpha_\infty} x_\alpha$, where $x_\alpha \in X_\alpha$ for all α sufficiently close to α_∞. The ultraproduct $\prod_{\alpha \to \alpha_\infty} X$ of a single set X is the *ultrapower* of X and is denoted *X.

Ultralimits of real numbers (i.e., elements of *$\mathbf{R}$) will be called *limit real numbers*; similarly one defines limit natural numbers, limit complex numbers, etc. Ordinary numbers will be called *standard* numbers to distinguish them from limit numbers, thus for instance a limit real number is an ultralimit of standard real numbers. All the usual arithmetic operations and relations on standard numbers are inherited by their limit analogues; for instance, a limit real number $\lim_{\alpha \to \alpha_\infty} x_\alpha$ is larger than another $\lim_{\alpha \to \alpha_\infty} y_\alpha$ if one has $x_\alpha > y_\alpha$ for all α sufficiently close to α_∞. The axioms of a non-principal ultrafilter ensure that these relations and operations on limit numbers obey the same axioms as their standard counterparts[3].

Ultraproducts of sets will be called *limit sets*; they are roughly analogous to "elementary sets" in measure theory. Ultraproducts of finite sets will be called *limit finite sets*. Thus, for instance, if $N = \lim_{\alpha \to \alpha_\infty} N_\alpha$ is a limit natural number, then $[N] = \prod_{\alpha \to \alpha_\infty} [N_\alpha]$ is a limit finite set, and can be identified with the set of limit natural numbers between 1 and N.

Remark 1.1.22. In the language of *non-standard analysis*, limit numbers and limit sets are known as *non-standard numbers* and *internal sets*, respectively. We will, however, use the language of ultralimit analysis rather than non-standard analysis in order to emphasise the fact that limit objects are the ultralimits of standard objects; see Section 2.1 for further discussion of this perspective.

---

[3]The formalisation of this principle is *Los's theorem*, which roughly speaking asserts that any first-order sentence which is true for standard objects, is also true for their limit counterparts.

1.1. Equidistribution in tori

Given a sequence of functions $f_\alpha: X_\alpha \to Y_\alpha$, we can form the *ultralimit* $\lim_{\alpha \to \alpha_\infty} f_\alpha: \lim_{\alpha \to \alpha_\infty} X_\alpha \to \lim_{\alpha \to \alpha_\infty} Y_\alpha$ by the formula

$$\left(\lim_{\alpha \to \alpha_\infty} f_\alpha\right)\left(\lim_{\alpha \to \alpha_\infty} x_\alpha\right) := \lim_{\alpha \to \alpha_\infty} f_\alpha(x_\alpha);$$

one easily verifies that this is a well-defined function between the two ultraproducts. We refer to ultralimits of functions as *limit functions*; they are roughly analogous to "simple functions" in measurable theory. We identify every standard function $f: X \to Y$ with its ultralimit $\lim_{\alpha \to \alpha_\infty} f: {}^*X \to {}^*Y$, which extends the original function f.

Now we introduce limit asymptotic notation, which is deliberately chosen to be similar (though not identical) to ordinary asymptotic notation. Given two limit numbers X, Y, we write $X \ll Y$, $Y \gg X$, or $X = O(Y)$ if we have $|X| \leq CY$ for some standard $C > 0$. We also write $X = o(Y)$ if we have $|X| \leq cY$ for every standard $c > 0$; thus for any limit numbers X, Y with $Y > 0$, exactly one of $|X| \gg Y$ and $X = o(Y)$ is true. A limit real is said to be *bounded* if it is of the form $O(1)$, and *infinitesimal* if it is of the form $o(1)$; similarly for limit complex numbers. Note that the bounded limit reals are a subring of the limit reals, and the infinitesimal limit reals are an ideal of the bounded limit reals.

Exercise 1.1.26 (Relation between limit asymptotic notation and ordinary asymptotic notation). Let $X = \lim_{\alpha \to \alpha_\infty} X_\alpha$ and $Y = \lim_{\alpha \to \alpha_\infty} Y_\alpha$ be two limit numbers.

(i) Show that $X \ll Y$ if and only if there exists a standard $C > 0$ such that $|X_\alpha| \leq CY_\alpha$ for all α sufficiently close to α_0.

(ii) Show that $X = o(Y)$ if and only if, for every standard $\varepsilon > 0$, one has $|X_\alpha| \leq \varepsilon Y_\alpha$ for all α sufficiently close to α_0.

Exercise 1.1.27. Show that every bounded limit real number x has a unique decomposition $x = \text{st}(x) + (x - \text{st}(x))$, where $\text{st}(x)$ is a standard real (called the *standard part* of x) and $x - \text{st}(x)$ is infinitesimal.

We now give the analogue of single-scale equidistribution in the ultralimit setting.

Definition 1.1.23 (Ultralimit equidistribution). Let $X = (X, d)$ be a standard compact metric space, let N be an unbounded limit natural number, and let $x: [N] \to {}^*X$ be a limit function. We say that x is *equidistributed* with respect to a (standard) Borel probability measure μ on X if one has

$$\text{st} \mathbf{E}_{n \in [N]} f(x(n)) = \int_X f \, d\mu$$

for all standard continuous functions $f \in C(X)$. Here, we define the expectation of a limit function in the obvious limit manner, thus

$$\mathbf{E}_{n \in [N]} f(x(n)) = \lim_{\alpha \to \alpha_\infty} \mathbf{E}_{n \in [N_\alpha]} f(x_\alpha(n))$$

if $N = \lim_{\alpha \to \alpha_\infty} N_\alpha$ and $x = \lim_{\alpha \to \alpha_\infty} x_\alpha$.

We say that x is *totally equidistributed* relative to μ if the sequence $n \mapsto x(qn + r)$ is equidistributed on $[N/q]$ for every standard $q > 0$ and $r \in \mathbf{Z}$ (extending x arbitrarily outside $[N]$ if necessary).

Remark 1.1.24. One could just as easily replace the space of continuous functions by any dense subclass in the uniform topology, such as the space of Lipschitz functions.

The ultralimit notion of equidistribution is closely related to that of both asymptotic equidistribution and single-scale equidistribution, as the following exercises indicate:

Exercise 1.1.28 (Asymptotic equidistribution vs. ultralimit equidistribution). Let $x \colon \mathbf{N} \to X$ be a sequence into a standard compact metric space (which can then be extended from a map from $^*\mathbf{N}$ to $^*X$ as usual), let μ be a Borel probability measure on X. Show that x is asymptotically equidistributed on $\mathbf{N}$ with respect to μ if and only if x is equidistributed on $[N]$ for every unbounded natural number N and every choice of non-principal ultrafilter α_∞.

Exercise 1.1.29 (Single-scale equidistribution vs. ultralimit equidistribution). For every $\alpha \in \mathbf{N}$, let N_α be a natural number that goes to infinity as $\alpha \to \infty$, let $x_\alpha \colon [N_\alpha] \to X$ be a map to a standard compact metric space. Let μ be a Borel probability measure on X. Write $N := \lim_{\alpha \to \alpha_\infty} N_\alpha$ and $x := \lim_{\alpha \to \alpha_\infty} x_\alpha$ for the ultralimits. Show that x is equidistributed with respect to μ if and only if, for every standard $\delta > 0$, x_α is δ-equidistributed with respect to μ for all α sufficiently close to α_∞.

In view of these correspondences, it is thus not surprising that one has ultralimit analogues of the asymptotic and single-scale theory. These analogues tend to be *logically equivalent* to the single-scale counterparts (once one concedes all quantitative bounds), but are *formally similar* (though not identical) to the asymptotic counterparts, thus providing a bridge between the two theories, which we can summarise by the following three statements:

(i) Asymptotic theory is analogous to ultralimit theory (in particular, the statements and proofs are formally similar);

(ii) ultralimit theory is logically equivalent to qualitative finitary theory; and

1.1. Equidistribution in tori

(iii) quantitative finitary theory is a strengthening of qualitative finitary theory.

For instance, here is the ultralimit version of the Weyl criterion:

Exercise 1.1.30 (Ultralimit Weyl equidistribution criterion). Let $x\colon [N] \to {}*\mathbf{T}^d$ be a limit function for some unbounded N and standard d. Then x is equidistributed if and only if

$$\mathbf{E}_{n\in[N]} e(k \cdot x(n)) = o(1) \tag{1.7}$$

for all standard $k \in \mathbf{Z}^d \backslash \{0\}$. *Hint:* Mimic the proof of Proposition 1.1.2.

Exercise 1.1.31. Use Exercise 1.1.30 to recover a weak version of Proposition 1.1.13, in which the quantities δ^{cd}, δ^{C_d} are replaced by (ineffective) functions of δ that decay to zero as $\delta \to 0$. Conversely, use this weak version to recover Exercise 1.1.30. (*Hint:* Similar arguments appear in Section 2.1.)

Exercise 1.1.32. With the notation of Exercise 1.1.30, show that x is totally equidistributed if and only if

$$\mathbf{E}_{n\in[N]} e(k \cdot x(n)) e(\theta n) = o(1)$$

for all standard $k \in \mathbf{Z}^d \backslash \{0\}$ and standard rational θ.

Exercise 1.1.33. With the notation of Exercise 1.1.30, show that x is equidistributed in $\mathbf{T}^d$ on $[N]$ if and only if $k \cdot x$ is equidistributed in $\mathbf{T}$ on $[N]$ for every non-zero standard $k \in \mathbf{Z}^d$.

Now we establish the ultralimit version of the linear equidistribution criterion:

Exercise 1.1.34. Let $\alpha, \beta \in {}*\mathbf{T}^d$, and let N be an unbounded integer. Show that the following are equivalent:

(i) The sequence $n \mapsto n\alpha + \beta$ is equidistributed on $[N]$.

(ii) The sequence $n \mapsto n\alpha + \beta$ is totally equidistributed on $[N]$.

(iii) α is *irrational to scale* $1/N$, in the sense that $k \cdot \alpha \neq O(1/N)$ for any non-zero standard $k \in \mathbf{Z}^d$.

Note that in the ultralimit setting, assertions such as $k \cdot \alpha \neq O(1/N)$ make perfectly rigorous sense (it means that $|k \cdot \alpha| \geq C/N$ for every standard C), but when using finitary asymptotic big-O notation

Next, we establish the analogue of the van der Corput lemma:

Exercise 1.1.35 (van der Corput lemma, ultralimit version). Let N be an unbounded integer, and let $x\colon [N] \to {}*\mathbf{T}^d$ be a limit sequence. Let $H = o(N)$ be unbounded, and suppose that the derivative sequence $\partial_h x \colon n \mapsto$

$x(n+h) - x(n)$ is equidistributed on $[N]$ for $\gg H$ values of $h \in [H]$ (by extending x arbitrarily outside of $[N]$). Show that x is equidistributed on $[N]$. Similarly, "equidistributed" is replaced by "totally equidistributed".

Here is the analogue of the Vinogradov lemma:

Exercise 1.1.36 (Vinogradov lemma, ultralimit version). Let $\alpha \in {*}\mathbf{T}$, N be unbounded, and $\varepsilon > 0$ be infinitesimal. Suppose that $\|n\alpha\|_\mathbf{T} \leq \varepsilon$ for $\gg N$ values of $n \in [-N, N]$. Show that there exists a positive standard integer q such that $\|\alpha q\|_\mathbf{T} \ll \varepsilon/N$.

These two lemmas allow us to establish the ultralimit polynomial equidistribution theory:

Exercise 1.1.37. Let $P \colon {*}\mathbf{Z} \to {*}\mathbf{T}^d$ be a polynomial sequence $P(n) := \alpha_s n^s + \cdots + \alpha_0$ with s, d standard, and $\alpha_0, \ldots, \alpha_s \in {*}\mathbf{T}^d$. Let N be an unbounded natural number. Suppose that P is not totally equidistributed on $[N]$. Show that there exists a non-zero standard $k \in \mathbf{Z}^d$ with $\|k \cdot \alpha_s\|_\mathbf{T} \ll N^{-s}$.

Exercise 1.1.38. With the hypotheses of Exercise 1.1.37, show in fact that there exists a non-zero standard $k \in \mathbf{Z}^d$ such that $\|k \cdot \alpha_i\|_\mathbf{T} \ll N^{-i}$ for all $i = 0, \ldots, s$.

Exercise 1.1.39 (Ultralimit equidistribution theorem for abelian polynomial sequences). Let P be a polynomial map from ${*}\mathbf{Z}$ to ${*}\mathbf{T}^d$ of some standard degree $s \geq 0$. Let N be an unbounded natural number. Then there exists a decomposition
$$P = P_{\text{smth}} + P_{\text{equi}} + P_{\text{rat}}$$
into polynomials of degree s, where

(i) (P_{smth} is smooth) The ith coefficient $\alpha_{i,\text{smth}}$ of P_{smth} has size $O(N^{-i})$. In particular, on the interval $[N]$, P_{smth} is Lipschitz with homogeneous norm $O(1/N)$.

(ii) (P_{equi} is equidistributed) There exists a standard subtorus T of $\mathbf{T}^d$, such that P_{equi} takes values in T and is totally equidistributed on $[N]$ in this torus.

(iii) (P_{rat} is rational) The coefficients $\alpha_{i,\text{rat}}$ of P_{rat} are standard rational elements of $\mathbf{T}^d$. In particular, there is a standard positive integer q such that $qP_{\text{rat}} = 0$ and P_{rat} is periodic with period q.

Exercise 1.1.40. Show that the torus T is uniquely determined by P, and decomposition $P = P_{\text{smth}} + P_{\text{equi}} + P_{\text{rat}}$ in Exercise 1.1.39 is unique up to expressions taking values in T (i.e., if one is given another decomposition

1.1. Equidistribution in tori

$P = P'_{\text{smth}} + P'_{\text{equi}}, P'_{\text{rat}}$, then P_i and P'_i differ by expressions taking values in T).

Exercise 1.1.41 (Recurrence). Let P be a polynomial map from $*\mathbf{Z}$ to $*\mathbf{T}^d$ of some standard degree s, and let N be an unbounded natural number. Show that for every standard $\varepsilon > 0$ and every $n_0 \in N$, we have

$$|\{n \in [N] : \|P(n) - P(n_0)\| \leq \varepsilon\}| \gg N$$

and more generally

$$|\{r \in [-N, N] : \|P(n_0 + jr) - P(n_0)\| \leq \varepsilon \text{ for } j = 0, 1, \ldots, k-1\}| \gg N$$

for any standard k.

As before, there are also multidimensional analogues of this theory. We shall just state the main results without proof:

Definition 1.1.25 (Multidimensional equidistribution). Let X be a standard compact metric space, let N be an unbounded limit natural number, let $m \geq 1$ be standard, and let $x \colon [N]^m \to *X$ be a limit function. We say that x is *equidistributed* with respect to a (standard) Borel probability measure μ on X if one has

$$\text{st}\mathbf{E}_{n \in [N]^m} 1_B(n/N) f(x(n)) = \text{mes}(\Omega) \int_X f \, d\mu$$

for every standard box $B \subset [0,1]^m$ and for all standard continuous functions $f \in C(X)$.

We say that x is *totally equidistributed* relative to μ if the sequence $n \mapsto x(qn + r)$ is equidistributed on $[N/q]^d$ for every standard $q > 0$ and $r \in \mathbf{Z}^m$ (extending x arbitrarily outside $[N]$ if necessary).

Remark 1.1.26. One can replace the indicators 1_B by many other classes, such as indicators of standard convex sets, or standard open sets whose boundary has measure zero, or continuous or Lipschitz functions.

Theorem 1.1.27 (Multidimensional ultralimit equidistribution theorem for abelian polynomial sequences). *Let $m, d, s \geq 0$ be standard integers, and let P be a polynomial map from $*\mathbf{Z}^m$ to $*\mathbf{T}^d$ of degree s. Let N be an unbounded natural number. Then there exists a decomposition*

$$P = P_{\text{smth}} + P_{\text{equi}} + P_{\text{rat}}$$

into polynomials of degree s, where

  (i) *(P_{smth} is smooth) The ith coefficient $\alpha_{i,\text{smth}}$ of P_{smth} has size $O(N^{-|i|})$ for every multi-index $i = (i_1, \ldots, i_m)$. In particular, on the interval $[N]$, P_{smth} is Lipschitz with homogeneous norm $O(1/N)$.*

(ii) (P_{equi} is equidistributed) There exists a standard subtorus T of $\mathbf{T}^d$, such that P_{equi} takes values in T and is totally equidistributed on $[N]^m$ in this torus.

(iii) (P_{rat} is rational) The coefficients $\alpha_{i,\text{rat}}$ of P_{rat} are standard rational elements of $\mathbf{T}^d$. In particular, there is a standard positive integer q such that $qP_{\text{rat}} = 0$ and P_{rat} is periodic with period q.

Proof. This is implicitly in [**GrTa2011**]; the result is phrased using the language of single-scale equidistribution, but this easily implies the ultralimit version. $\square$

1.2. Roth's theorem

We now give a basic application of Fourier analysis to the problem of counting additive patterns in sets, namely the following famous theorem of Roth [**Ro1953**]:

Theorem 1.2.1 (Roth's theorem). *Let A be a subset of the integers $\mathbf{Z}$ whose upper density*
$$\overline{\delta}(A) := \limsup_{N \to \infty} \frac{|A \cap [-N, N]|}{2N + 1}$$
is positive. Then A contains infinitely many arithmetic progressions $a, a + r, a + 2r$ of length three, with $a \in \mathbf{Z}$ and $r > 0$.

This is the first non-trivial case of *Szemerédi's theorem* [**Sz1975**], which is the same assertion but with length three arithmetic progressions replaced by progressions of length k for any k.

As it turns out, one can prove Roth's theorem by an application of linear Fourier analysis by comparing the set A (or more precisely, the indicator function 1_A of that set, or of pieces of that set) against linear characters $n \mapsto e(\alpha n)$ for various frequencies $\alpha \in \mathbf{R}/\mathbf{Z}$. There are two extreme cases to consider (which are model examples of a more general dichotomy between structure and randomness, as discussed in [**Ta2008**]). One is when A is aligned almost completely with one of these linear characters, for instance, by being a *Bohr set* of the form
$$\{n \in \mathbf{Z} : \|\alpha n - \theta\|_{\mathbf{R}/\mathbf{Z}} < \varepsilon\}$$
or, more generally, of the form
$$\{n \in \mathbf{Z} : \alpha n \in U\}$$
for some multi-dimensional frequency $\alpha \in \mathbf{T}^d$ and some open set U. In this case, arithmetic progressions can be located using the equidistribution theory from Section 1.1. At the other extreme, one has *Fourier-uniform* or *Fourier-pseudorandom sets*, whose correlation with any linear character is

1.2. Roth's theorem

negligible. In this case, arithmetic progressions can be produced in abundance via a Fourier-analytic calculation.

To handle the general case, one must somehow synthesise together the argument that deals with the structured case with the argument that deals with the random case. There are several known ways to do this, but they can be basically classified into two general methods, namely the *density increment argument* (or L^∞ increment argument) and the *energy increment argument* (or L^2 increment argument).

The idea behind the density increment argument is to introduce a dichotomy: either the object A being studied is pseudorandom (in which case one is done), or else one can use the theory of the structured objects to locate a sub-object of significantly higher "density" than the original object. As the density cannot exceed one, one should thus be done after a finite number of iterations of this dichotomy. This argument was introduced by Roth in his original proof [**Ro1953**] of the above theorem.

The idea behind the energy increment argument is instead to decompose the original object A into two pieces (and, sometimes, a small additional error term): a *structured component* that captures all the structured objects that have significant correlation with A, and a *pseudorandom component* which has no significant correlation with any structured object. This decomposition usually proceeds by trying to maximise the "energy" (or L^2 norm) of the structured component, or dually by trying to minimise the energy of the residual between the original object and the structured object. This argument appears for instance in the proof of the *Szemerédi regularity lemma* [**Sz1978**] (which, not coincidentally, can also be used to prove Roth's theorem), and is also implicit in the ergodic theory approach to such problems (through the machinery of conditional expectation relative to a factor, which is a type of orthogonal projection, the existence of which is usually established via an energy increment argument). However, one can also deploy the energy increment argument in the Fourier analytic setting, to give an alternate Fourier-analytic proof of Roth's theorem that differs in some ways from the density increment proof.

In this section we give both Fourier-analytic proofs of Roth's theorem, one proceeding via the density increment argument, and the other by the energy increment argument. As it turns out, both of these arguments extend to establish Szemerédi's theorem, and more generally in counting other types of patterns, but this is non-trivial (requiring some sort of *inverse conjecture* for the Gowers uniformity norms in both cases); we will discuss this further in later sections.

1.2.1. The density increment argument. We begin with the density increment argument. We first rephrase Roth's theorem in a finitary form:

Theorem 1.2.2 (Roth's theorem, again). *For every $\delta > 0$, there exists an $N_0 = N_0(\delta) > 0$, such that for every $N \geq N_0$, and every $A \subset [N]$ with $|A| \geq \delta N$, A contains an arithmetic progression of length three.*

Exercise 1.2.1. Show that Theorem 1.2.1 and Theorem 1.2.2 are equivalent.

We prove Theorem 1.2.2 by a downward induction on the density parameter δ. Let $P(\delta)$ denote the proposition that Theorem 1.2.2 holds for that value of δ (i.e., for sufficiently large N and all $A \subset [N]$ with $|A| \geq \delta N$, A contains an arithmetic progression of length three). Our objective is to show that $P(\delta)$ holds for all $\delta > 0$.

Clearly, $P(\delta)$ is (vacuously) true for $\delta > 1$ (and trivially true for $\delta \geq 1$). It is also monotone in the sense that if $P(\delta)$ holds for some δ, then $P(\delta')$ holds for all $\delta' > \delta$. To downwardly induct on δ, we will prove the following dichotomy:

Proposition 1.2.3 (Lack of progressions implies density increment). *Let $\delta > 0$, let N be sufficiently large depending on δ, and let $A \subset [N]$ be such that $|A| \geq \delta N$. Then one of the following holds:*

  (i) *A contains an arithmetic progression of length three; or*

  (ii) *there exists a subprogression P of $[N]$ of length at least N' such that $|A \cap P| \geq (\delta + c(\delta))|P|$, where $N' = N'(N)$ goes to infinity as $N \to \infty$, and $c(\delta) > 0$ is bounded away from zero whenever δ is bounded away from zero.*

Let us see why Proposition 1.2.3 implies Theorem 1.2.2. It is slightly more convenient to use a "well-ordering principle" argument rather than an induction argument, though unsurprisingly the two approaches turn out to be equivalent. Let δ_* be the infimum of all δ for which $P(\delta)$ holds, thus $0 \leq \delta_* \leq 1$. If $\delta_* = 0$, then we are done, so suppose that δ_* is non-zero. Then for any $\varepsilon > 0$, $P(\delta_* - \varepsilon)$ is false, thus there exist arbitrarily large N and $A \subset [N]$ with $|A| \geq (\delta_* - \varepsilon)N$ with no progressions of length three. By Proposition 1.2.3, we can thus find a subprogression P of N of length at least N' with $|A \cap P| \geq (\delta_* - \varepsilon + c(\delta_* - \varepsilon))|P|$; if ε is small enough, this implies that $|A \cap P| \geq (\delta_* + \varepsilon)|P|$. We then use an affine transformation to map P to $[N']$ (noting crucially that the property of having no arithmetic progressions of a given length is preserved by affine transformations). As N can be arbitrarily large, N' can be arbitrarily large also. Since $P(\delta_* + \varepsilon)$ is true, we see that $A \cap P$ contains an arithmetic progression of length three, hence A does also; which gives the desired contradiction.

1.2. Roth's theorem

It remains to prove Proposition 1.2.3. There are two main steps. The first relies heavily on the fact that the progressions only have length three, and is proven via Fourier analysis:

Proposition 1.2.4 (Lack of progressions implies correlation with a linear phase). *Let $\delta > 0$, let N be sufficiently large depending on δ, let $A \subset [N]$ be such that $|A| = \delta' N$ for some $\delta' \geq \delta$, with A containing no arithmetic progressions of length three. Then there exists $\alpha \in \mathbf{R}/\mathbf{Z}$ such that $|\mathbf{E}_{n \in [N]}(1_A(n) - \delta')e(-\alpha n)| \gg \delta^2$.*

Proof. In order to use Fourier analysis, it will be convenient to embed $[N]$ inside a cyclic group $\mathbf{Z}/N'\mathbf{Z}$, where N' is equal to (say) $2N + 1$; the exact choice here is only of minor importance, though it will be convenient to take N' to be odd. We introduce the trilinear form

$$\Lambda(f, g, h) := \mathbf{E}_{n,r \in \mathbf{Z}/N'\mathbf{Z}} f(n)g(n+r)h(n+2r)$$

for any functions $f, g, h \colon \mathbf{Z}/N'\mathbf{Z} \to \mathbf{C}$; we then observe that the quantity

$$\Lambda(1_A, 1_A, 1_A) = \mathbf{E}_{n,r \in \mathbf{Z}/N'\mathbf{Z}} 1_A(n) 1_A(n+r) 1_A(n+2r)$$

(extending 1_A by zero outside of $[N]$) is equal to the number of arithmetic progressions $n, n+r, n+2r$ in A (counting the degenerate progressions in which $r = 0$, and also allowing for r to be negative), divided by the normalising factor of $(N')^2$. On the other hand, by hypothesis, A contains no non-degenerate arithmetic progressions of length three, and clearly has $|A| \leq N$ degenerate progressions; thus we have

$$(1.8) \qquad \Lambda(1_A, 1_A, 1_A) \ll 1/N.$$

On the other hand, from the Fourier inversion formula on the cyclic group $\mathbf{Z}/N'\mathbf{Z}$ we may write

$$f(n) = \sum_{\alpha \in \frac{1}{N'}\mathbf{Z}/\mathbf{Z}} \hat{f}(\alpha) e(\alpha n)$$

for any function $f \colon \mathbf{Z}/N'\mathbf{Z} \to \mathbf{C}$, where $\hat{f}(\alpha)$ are the Fourier coefficients

$$\hat{f}(\alpha) := \mathbf{E}_{n \in \mathbf{Z}/N'\mathbf{Z}} f(n) e(-\alpha n).$$

We may thus write $\Lambda(f, g, h)$ as

$$\sum_{\alpha_1, \alpha_2, \alpha_3 \in \frac{1}{N'}\mathbf{Z}/\mathbf{Z}} \hat{f}(\alpha_1) \hat{g}(\alpha_2) \hat{h}(\alpha_3)$$

$$(1.9) \qquad \mathbf{E}_{n,r \in \mathbf{Z}/N'\mathbf{Z}} e(\alpha_1 n + \alpha_2(n+r) + \alpha_3(n+2r)).$$

Now observe that we have the identity

$$\alpha n - 2\alpha(n+r) + \alpha(n+2r) = 0,$$

so the phase $\alpha_1 n + \alpha_2(n+r) + \alpha_3(n+2r)$ is trivial whenever $(\alpha_1, \alpha_2, \alpha_3)$ is of the form $(\alpha, -2\alpha, \alpha)$, and so the expectation in (1.9) is equal to 1. Conversely, if $(\alpha_1, \alpha_2, \alpha_3)$ is not of this form, then the phase is non-trivial, and from Fourier analysis we conclude that the expectation in (1.9) vanishes. We conclude that the left-hand side of (1.8) can be expressed as

$$\sum_{\alpha \in \frac{1}{N'}\mathbf{Z}/\mathbf{Z}} \hat{f}(\alpha)\hat{g}(-2\alpha)\hat{h}(\alpha).$$

Now using Plancherel's theorem we have

$$\sum_{\alpha \in \frac{1}{N'}\mathbf{Z}/\mathbf{Z}} |\hat{f}(\alpha)|^2 = \|f\|^2_{L^2(\mathbf{Z}/N'\mathbf{Z})}$$

(using normalised counting measure). Using this and Hölder's inequality (and the fact that N' is odd), we obtain the bounds

$$(1.10) \qquad |\Lambda(f,g,h)| \leq \|f\|_{L^2(\mathbf{Z}/N'\mathbf{Z})} \|g\|_{L^2(\mathbf{Z}/N'\mathbf{Z})} \sup_{\xi \in \mathbf{Z}/N'\mathbf{Z}} |\hat{h}(\xi)|$$

and similarly for permutations of f, g, h on the right-hand side.

We could apply this directly to $\Lambda(1_A, 1_A, 1_A)$, but this is not useful, since we seek a *lower* bound on this quantity rather than an upper bound. To get such a lower bound, we split $1_A = \delta' 1_{[N]} + f$, where $f := 1_A - \delta' 1_{[N]}$ is the mean zero portion of 1_A, and use trilinearity to split $\Lambda(1_A, 1_A, 1_A)$ into a main term $\Lambda(\delta' 1_{[N]}, \delta' 1_{[N]}, \delta' 1_{[N]})$, plus seven other error terms involving $1_A = \delta' 1_{[N]}$ and f, with each error term involving at least one copy of f. The main term can be computed explicitly as

$$\Lambda(\delta' 1_{[N]}, \delta' 1_{[N]}, \delta' 1_{[N]}) \gg \delta^3.$$

Comparing this with (1.8), we conclude that one of the error terms must have magnitude $\gg \delta^3$ also. For sake of concreteness, let us say that

$$|\Lambda(f, \delta' 1_{[N]}, f)| \gg \delta^3;$$

the other cases are similar and are left to the reader.

From the triangle inequality we see that $f, \delta' 1_{[N]}$ have an $L^2(\mathbf{Z}/N'\mathbf{Z})$ norm of $O(\delta^{1/2})$, and so from (1.10) one has

$$|\Lambda(f, \delta' 1_{[N]}, f)| \ll \delta \sup_{\xi \in \mathbf{Z}/N'\mathbf{Z}} |\hat{f}(\xi)|,$$

and so we conclude that

$$|\hat{f}(\xi)| \gg \delta^2$$

for some $\xi \in \mathbf{Z}/N'\mathbf{Z}$. Similarly for other error terms, though sometimes one will need a permutation of (1.10) instead of (1.10) itself. The claim follows. □

1.2. Roth's theorem

Remark 1.2.5. The above argument relied heavily on the fact that there was only a one-parameter family of linear relations between $n, n+r, n+2r$. The same argument does not work directly for finding arithmetic progressions of length longer than three; we return to this point in later sections.

The second step converts correlation with a linear character into a density increment on a subprogression:

Proposition 1.2.6 (Fragmenting a linear character into progressions). *Let $N \geq 1$, let $\varepsilon > 0$, and let $\phi(n) := e(\alpha n)$ be a linear phase. Then there exists $N' = N'(N, \varepsilon)$ which goes to infinity as $N \to \infty$ for fixed ε, and a partition*

$$[N] = \bigcup_{j=1}^{J} P_j \cup E$$

of $[N]$ into arithmetic progressions P_j of length at least N', together with an error term E of cardinality at most $O(\varepsilon N)$, such that ϕ fluctuates by at most $O(\varepsilon)$ on each progression P_j (i.e., $|\phi(x) - \phi(y)| \ll \varepsilon$ whenever $x, y \in P_j$).

Proof. We may assume that N is sufficiently large depending on ε, as the claim is trivial otherwise (just set $N' = 1$).

Fix ε, and let N' be a slowly growing function of N to be chosen later. By using recurrence for the linear phase $n \mapsto \alpha n$, we can find a shift $h \geq 1$ of size $h = O_{N', \varepsilon}(1)$ such that $\|\alpha h\|_{\mathbf{R}/\mathbf{Z}} \leq \varepsilon/N'$. We then partition $[N]$ into h arithmetic progressions of spacing h, and then partition each of those progressions in turn into subprogressions P_j of spacing h and length N', plus an error of cardinality at most N', leading to an error set E of cardinality at most $hN' = O_{N', \varepsilon}(1)$. On each of the P_j, αn fluctuates by at most ε. The claim then follows by choosing N' to be a sufficiently slowly growing function of N. $\square$

Now we can prove Proposition 1.2.3 (and thus Roth's theorem). Let N, δ, δ', A be as in Proposition 1.2.3. By Proposition 1.2.4 (if N is large enough), we can find α for which

$$|\mathbf{E}_{n \in [N]}(1_A(n) - \delta')e(-\alpha n)| \gg \delta^2.$$

We now let $\varepsilon > 0$ be a small quantity depending on δ to be chosen later (actually it turns out that we can take ε to be a small multiple of δ^2) and apply Proposition 1.2.6 to decompose $[N]$ into progressions $P_1, \ldots, P_J$ and an error term E with the stated properties. Then we have

$$\mathbf{E}_{n \in [N]}(1_A(n) - \delta')e(-\alpha n) = \frac{1}{N}(\sum_{j=1}^{J} \sum_{n \in P_j}(1_A(n) - \delta')e(-\alpha n)) + O(\varepsilon).$$

Since $e(-\alpha n)$ fluctuates by at most ε on P_j, we can apply the triangle inequality and conclude that

$$|\mathbf{E}_{n\in[N]}(1_A(n) - \delta')e(-\alpha n)| \leq \frac{1}{N}\left|\sum_{j=1}^{J}\sum_{n\in P_j}(1_A(n) - \delta')\right| + O(\varepsilon).$$

If ε is sufficiently small depending on δ, we conclude that

(1.11) $$\sum_{j=1}^{J}|\sum_{n\in P_j}(1_A(n) - \delta')| \gg \delta^2 N.$$

On the other hand, as δ' is the mean of 1_A on $[N]$, we have

$$\sum_{n\in[N]}(1_A(n) - \delta') = 0$$

and thus

$$\sum_{j=1}^{J}\sum_{n\in P_j}(1_A(n) - \delta') = O(\varepsilon).$$

Adding this to (1.11) and noting that $|x| + x = 2\max(x, 0)$ for real x, we conclude (for ε small enough) that

$$\sum_{j=1}^{J}\max(\sum_{n\in P_j}(1_A(n) - \delta'), 0) \gg \delta^2 N$$

and hence by the pigeonhole principle we can find j such that

$$\max(\sum_{n\in P_j}(1_A(n) - \delta'), 0) \gg \delta^2|P_j|$$

or, in other words,

$$|A \cap P_j|/|P_j| \geq \delta' + c\delta^2$$

for some absolute constant $c > 0$, and Proposition 1.2.3 follows.

It is possible to rewrite the above argument in the ultralimit setting, though it only makes the argument slightly shorter as a consequence. We sketch this alternate formulation below.

Exercise 1.2.2. Let δ_* be as above.

(i) Show that if N is an unbounded limit natural number, and $A \subset [N]$ is a limit subset whose density $\text{st}(|A|/N)$ is strictly greater than δ_*, then A contains a (limit) arithmetic progression $n, n+r, n+2r$ of length three (with $r \neq 0$).

(ii) Show that there exists an unbounded limit natural number N and a limit subset $A \subset [N]$ of density $\text{st}(|A|/N) = \delta_*$, which does not contain any arithmetic progressions of length three.

Exercise 1.2.3. Show that if N is an unbounded limit natural number, and $A \subset [N]$ is a limit subset of positive density (st $|A|/N) = \delta' > 0$ with no arithmetic progressions of length three, then there exists a limit real α such that $|\mathbf{E}_{n \in [N]}(1_A(n) - \delta')e(-\alpha n)| \gg 1$.

Exercise 1.2.4. If N is an unbounded limit natural number, and α is a limit real, show that one can partition $[N] = \bigcup_{j=1}^{J} P_j \cup E$, where J is a limit natural number, the P_j are limit arithmetic subprogressions of $[N]$ of unbounded length (with the map $j \mapsto P_j$ being a limit function), such that $n \mapsto e(\alpha n)$ fluctuates by $o(1)$ on each P_j (uniformly in j), and $|E| = o(N)$.

Exercise 1.2.5. Use the previous three exercises to reprove Roth's theorem.

Exercise 1.2.6 (Roth's theorem in bounded characteristic). Let F be a finite field, let $\delta > 0$, and let V be a finite vector space. Show that if the dimension of V is sufficiently large depending on F, δ, and if $A \subset V$ is such that $|A| \geq \delta|V|$, then there exists $a, r \in V$ with $r \neq 0$ such that $a, a+r, a+2r \in A$. (*Hint:* Mimic the above arguments (either finitarily, or with ultralimits), using hyperplanes as a substitute for subprogressions.)

Exercise 1.2.7 (Roth's theorem in finite abelian groups). Let G be a finite abelian group, and let $\delta > 0$. Show that if $|G|$ is sufficiently large depending on δ, and $A \subset G$ is such that $|A| \geq \delta|G|$, then there exists $a, r \in V$ with $r \neq 0$ such that $a, a+r, a+2r \in A$. (*Hint:* If there is an element of G of large order, one can use Theorem 1.2.2 and the pigeonhole principle. If all elements have bounded order, one can instead use Exercise 1.2.6.) This result (as well as the special case in the preceding exercise) was first established by Meshulam [**Me1995**].

1.2.2. The energy increment argument. Now we turn to the energy increment approach. This approach requires a bit more machinery to set up, but ends up being quite flexible and powerful (for instance, it is the starting point for my theorem with Ben Green establishing arbitrarily long progressions in the primes, which we do not know how to establish via density increment arguments).

Instead of passing from $[N]$ to a subprogression, we now instead *coarsen* $[N]$ to some partition (or *factor*) of $[N]$, as follows. Define a *factor* of $[N]$ to be a σ-algebra of subsets $\mathcal{B}$ of $[N]$, or equivalently a partition of $[N]$ into disjoint *atoms* or *cells* (with the elements of $\mathcal{B}$ then being the arbitary unions of atoms). Given a function $f \colon [N] \to \mathbf{C}$ and a factor $\mathcal{B}$, we define the *conditional expectation* $\mathbf{E}(f|\mathcal{B}) \colon [N] \to \mathbf{C}$ to be the function whose value at a given point $x \in [N]$ is given by the formula

$$\mathbf{E}(f|\mathcal{B})(x) := \frac{1}{|\mathcal{B}(x)|} \sum_{y \in \mathcal{B}(x)} f(y),$$

where $\mathcal{B}(x)$ is the unique atom of $\mathcal{B}$ that contains x. One can view the map $f \mapsto \mathbf{E}(f|\mathcal{B})$ as the orthogonal projection from $L^2([N])$ to $L^2(\mathcal{B})$, where $L^2([N])$ is the space of functions $f\colon [N] \to \mathbf{C}$ with the inner product

$$\langle f, g \rangle_{L^2([N])} := \mathbf{E}_{n \in [N]} f(n)\overline{g(n)}$$

and $L^2(\mathcal{B})$ is the subspace of functions in $L^2([N])$ which are measurable with respect to $\mathcal{B}$, or equivalently are constant on each atom of $\mathcal{B}$.

We say that one factor $\mathcal{B}'$ *refines* another $\mathcal{B}$ if $\mathcal{B} \subset \mathcal{B}'$, or equivalently if every atom of $\mathcal{B}$ is a union of atoms of $\mathcal{B}'$, or if every atom of $\mathcal{B}'$ is contained in an atom of $\mathcal{B}'$, or equivalently again if $L^2(\mathcal{B}) \subset L^2(\mathcal{B}')$. Given two factors $\mathcal{B}, \mathcal{B}'$, one can define their *join* $\mathcal{B} \vee \mathcal{B}'$ to be their least common refinement, thus the atoms in $\mathcal{B} \vee \mathcal{B}'$ are the non-empty intersections of atoms in $\mathcal{B}$ with atoms in $\mathcal{B}'$.

The idea is to split a given function f in $L^2([N])$ (and specifically, an indicator function 1_A) into a projection $\mathbf{E}(f|\mathcal{B})$ onto a "structured factor" $\mathcal{B}$ to obtain a "structured component" $\mathbf{E}(f|\mathcal{B})$, together with a "pseudorandom component" $f - \mathbf{E}(f|\mathcal{B})$ that is essentially orthogonal to all structured functions. This decomposition is related to the classical decomposition of a vector in a Hilbert space into its orthogonal projection onto a closed subspace V, plus the complementary projection to the orthogonal complement $V^\perp$; we will see the relationship between the two decompositions later when we pass to the ultralimit.

We need to make the notion of "structured" more precise. We begin with some definitions. We say that a function $f\colon [N] \to \mathbf{C}$ has *Fourier complexity* at most M if it can be expressed as

$$f(n) = \sum_{m=1}^{M'} c_m e(\alpha_m n)$$

for some $M' \leq M$ and some complex numbers $c_1, \ldots, c_{M'}$ of magnitude at most 1, and some real numbers $\alpha_1, \ldots, \alpha_{M'}$. Note that from the Fourier inversion formula that every function will have some finite Fourier complexity, but typically one expects the complexity to grow with N; only a few special functions will have complexity bounded uniformly in N. Also note that if f, g have Fourier complexity M, then $f+g, f-g, \overline{f}$, or fg all have Fourier complexity at most $O_M(1)$; informally[4], the space of bounded complexity functions forms an algebra.

Ideally, we would like to take "functions of bounded Fourier complexity" as our class of structured functions. For technical reasons (related to our desire to use indicator functions as structured functions), we need to take

---

[4]We will be able to formalise this statement after we take ultralimits.

1.2. Roth's theorem

an L^1 closure and work with the wider class of *Fourier measurable* functions as our structured class.

Definition 1.2.7 (Measurability). Let $\mathcal{F}\colon \mathbf{R}^+ \to \mathbf{R}^+$ be a function. We say that a function $f\colon [N] \to \mathbf{C}$ is *Fourier measurable* with growth function $\mathcal{F}$ if, for every $K > 1$, one can find a function $f_K\colon [N] \to \mathbf{C}$ of Fourier complexity at most $\mathcal{F}(K)$ such that $\mathbf{E}_{n \in [N]} |f(n) - f_K(n)| \leq 1/K$.

A subset A of $[N]$ is *Fourier measurable* with growth function $\mathcal{F}$ if 1_A is Fourier measurable with this growth function.

Exercise 1.2.8. Show that every interval $[a, b]$ in $[N]$ is Fourier measurable with some growth function $\mathcal{F}$ independent of N. (*Hint:* Apply *Fejér summation* to the Fourier series of $1_{[a,b]}$.)

Exercise 1.2.9. Let f be a Fourier-measurable function with some growth function $\mathcal{F}$, which is bounded in magnitude by A. Show that for every $K > 1$, one can find a function $\tilde{f}_K\colon [N] \to \mathbf{C}$ which also is bounded in magnitude by A, and of Fourier complexity $O_{A,\mathcal{F}(K)}(1)$, such that $\mathbf{E}_{n\in[N]} |f(n) - \tilde{f}_K(n)| \ll 1/K$. (*Hint:* Start with the approximating function f_K from Definition 1.2.7, which is already bounded in magnitude by $\mathcal{F}(K)$, and then set $\tilde{f}_K := P(f_K, \overline{f_K})$ where $P(z, \bar{z})$ is a polynomial bounded in magnitude by A on the ball of radius $\mathcal{F}(K)$ which is close to the identity function on the ball of radius A (such a function can be constructed via the *Stone-Weierstrass theorem*).)

Exercise 1.2.10. Show that if $f, g\colon [N] \to \mathbf{C}$ are bounded in magnitude by A, and are Fourier measurable with growth functions $\mathcal{F}$, then $f + g$, $\bar{f}$, and fg are Fourier measurable with some growth function $\mathcal{F}'$ depending only on A and $\mathcal{F}$.

Conclude that if $E, F \subset [N]$ are Fourier-measurable with growth function $\mathcal{F}$, then $[N] \backslash E$, $E \cup F$, and $E \cap F$ are Fourier-measurable with some growth function $\mathcal{F}'$ depending only on $\mathcal{F}$.

We thus see that Fourier-measurable sets morally[5] form a Boolean algebra.

Now we make a key observation (cf. [**ReTrTuVa2008**]):

Lemma 1.2.8 (Correlation with a Fourier character implies correlation with a Fourier-measurable set). Let $f\colon [N] \to \mathbf{C}$ be bounded in magnitude by 1, and suppose that $|\mathbf{E}_{n \in [N]} f(n) e(-\alpha n)| \geq \delta$ for some $\delta > 0$. Then there exists a Fourier-measurable set $E \subset [N]$ with some growth function $\mathcal{F}$ depending on δ, such that $|\mathbf{E}_{n \in [N]} f(n) 1_E(n)| \gg \delta$.

---

[5] Again, we can formalise this assertion once we pass to the ultralimit; we leave this formalisation to the interested reader.

Proof. By splitting f into real and imaginary parts, we may assume, without loss of generality, that f is real. Rotating $e(-\alpha n)$, we may find a real number θ such that

$$\mathbf{E}_{n\in[N]}f(n)\operatorname{Re} e(-\alpha n + \theta) \geq \delta.$$

We then express

$$\operatorname{Re} e(-\alpha n + \theta) = 1 - \int_{-1}^{1} 1_{E_t}(n)\, dt$$

where

$$E_t := \{n \in [N] : \operatorname{Re} e(-\alpha n + \theta) \leq t\}.$$

By Minkowski's inequality, we thus have either

$$|\mathbf{E}_{n\in[N]}f(n)| \geq \delta/2$$

or

$$\int_{-1}^{1} |\mathbf{E}_{n\in[N]}f(n)1_{E_t}(n)|\, dt \geq \delta/2.$$

In the former case we are done (setting $E = [N]$), so suppose that the latter holds. If all the E_t were uniformly Fourier-measurable, we would now be done in this case also by the pigeonhole principle. This is not quite true; however, it turns out that *most* E_t are uniformly measurable, and this will be enough. More precisely, let $\varepsilon > 0$ be a small parameter to be chosen later, and say that t is *good* if one has

$$|E_{t+r}\setminus E_{t-r}| \leq 2\varepsilon^{-1} rN$$

for all $r > 0$. Let $\Omega \subset [-1,1]$ be the set of all bad t. Observe that for each bad t, we have $M\mu(t) \geq \varepsilon^{-1}$, where μ is the probability measure

$$\mu(S) := \frac{1}{N}|\{n \in [N] : \operatorname{Re} e(-\alpha n + \theta) \in S\}|$$

and M is the Hardy-Littlewood maximal function

$$M\mu(t) := \sup_{r>0} \frac{1}{2r}\mu([t-r, t+r]).$$

Applying the *Hardy-Littlewood maximal inequality*

$$|\{t \in \mathbf{R} : M\mu(t) \geq \lambda\}| \ll \frac{1}{\lambda}\|\mu\|,$$

(see e.g. [**Ta2011**, §1.6] for a proof) we conclude that $|\Omega| \ll \varepsilon$. In particular, if ε is small enough compared with δ, we have

$$\int_{[-1,1]\setminus\Omega} |\mathbf{E}_{n\in[N]}f(n)1_{E_t}(n)|\, dt \gg \delta$$

and so by the pigeonhole principle, there exists a good t such that

$$|\mathbf{E}_{n\in[N]}f(n)1_{E_t}(n)| \gg \delta.$$

1.2. Roth's theorem

It remains to verify that E_t is good. For any $K > 0$, we have (as t is good) that
$$\mathbf{E}_{n \in [N]}(1_{E_{t+1/K}} - 1_{E_{t-1/K}}) \ll_\delta 1/K.$$
Applying *Urysohn's lemma*, we can thus find a smooth function $\eta \colon \mathbf{R} \to \mathbf{R}^+$ with $\eta(t') = 1$ for $t' < t - 1/K$ and $\eta(t') = 0$ for $t' > t + 1/K$ such that
$$\mathbf{E}_{n \in [N]} |1_{E_t}(n) - \eta(\operatorname{Re} e(-\alpha n + \theta))| \ll_\delta 1/K.$$
Using the *Weierstrass approximation theorem*, one can then approximate η uniformly by $O(1/K)$ on $[-1, 1]$ by a polynomial of degree $O_K(1)$ and coefficients $O_K(1)$. This allows one to approximate 1_{E_t} in L^1 norm to an accuracy of $O_\delta(1/K)$ by a function of Fourier complexity $O_K(1)$, and the claim follows. $\square$

Corollary 1.2.9 (Correlation implies energy increment). *Let $f \colon [N] \to [0,1]$, and let $\mathcal{B}$ be a factor generated by at most M atoms, each of which is Fourier-measurable with growth function $\mathcal{F}$. Suppose that we have the correlation*
$$|\langle f - \mathbf{E}(f|\mathcal{B}), e(\alpha \cdot)\rangle_{L^2([N])}| \geq \delta$$
for some $\delta > 0$ and $\alpha \in \mathbf{R}$. Then there exists a refinement $\mathcal{B}'$ generated by at most $2M$ atoms, each of which is Fourier-measurable with a growth function $\mathcal{F}'$ depending only on $\delta, \mathcal{F}$, such that
$$\|\mathbf{E}(f|\mathcal{B}')\|^2_{L^2([N])} - \|\mathbf{E}(f|\mathcal{B})\|^2_{L^2([N])} \gg \delta^2. \tag{1.12}$$

Proof. By Lemma 1.2.8, we can find a Fourier-measurable set E with some growth function $\mathcal{F}''$ depending on δ, such that
$$|\langle f - \mathbf{E}(f|\mathcal{B}), 1_E\rangle_{L^2([N])}| \gg \delta.$$
We let $\mathcal{B}'$ be the factor generated by $\mathcal{B}$ and E. As 1_E is measurable with respect to $\mathcal{B}'$, we may project onto $L^2(\mathcal{B}')$ and conclude that
$$|\langle \mathbf{E}(f|\mathcal{B}') - \mathbf{E}(f|\mathcal{B}), 1_E\rangle_{L^2([N])}| \gg \delta.$$
By Cauchy-Schwarz, we thus have
$$\|\mathbf{E}(f|\mathcal{B}') - \mathbf{E}(f|\mathcal{B})\|_{L^2([N])} \gg \delta.$$
Squaring and using Pythagoras' theorem, we obtain (1.12). The remaining claims in the corollary follow from Exercise 1.2.10. $\square$

We can then iterate this corollary via an *energy increment argument* to obtain

Proposition 1.2.10 (Weak arithmetic regularity lemma). *Let $f \colon [N] \to [0, 1]$, and let $\mathcal{B}$ be a factor generated by at most M atoms, each of which is Fourier-measurable with growth function $\mathcal{F}$. Let $\delta > 0$. Then there exists*

an extension $\mathcal{B}'$ of $\mathcal{B}$ generated by $O_{M,\delta}(1)$ atoms, each of which is Fourier-measurable with growth function $\mathcal{F}'$ depending on $\mathcal{F}, \delta$, such that

(1.13) $$|\langle f - \mathbf{E}(f|\mathcal{B}'), e(\alpha \cdot)\rangle_{L^2([N])}| < \delta$$

for all $\alpha \in \mathbf{R}$.

Proof. We initially set $\mathcal{B}'$ equal to $\mathcal{B}$. If (1.13) already holds, then we are done; otherwise, we invoke Corollary 1.2.9 to increase the "energy" $\|\mathbf{E}(f|\mathcal{B}')\|_{L^2}^2$ by $\gg \delta^2$, at the cost of possibly doubling the number of atoms in $\mathcal{B}'$, and also altering the growth function somewhat. We iterate this procedure; as the energy $\|\mathbf{E}(f|\mathcal{B}')\|_{L^2}^2$ is bounded between zero and one, and increases by $\gg \delta^2$ at each step, the procedure must terminate in $O(1/\delta^2)$ steps, at which point the claim follows. $\square$

It turns out that the power of this lemma is amplified if we iterate one more time, to obtain

Theorem 1.2.11 (Strong arithmetic regularity lemma). *Let $f\colon [N] \to [0,1]$, let $\varepsilon > 0$, and let $F\colon \mathbf{R}^+ \to \mathbf{R}^+$ be an arbitrary function. Then we can decompose $f = f_{\mathrm{str}} + f_{\mathrm{sml}} + f_{\mathrm{psd}}$ and find $1 \leq M = O_{\varepsilon, F}(1)$ such that*

(i) *(Nonnegativity) $f_{\mathrm{str}}, f_{\mathrm{str}} + f_{\mathrm{sml}}$ take values in $[0,1]$, and $f_{\mathrm{sml}}, f_{\mathrm{psd}}$ have mean zero;*

(ii) *(Structure) f_{str} is Fourier-measurable with a growth function $\mathcal{F}_M$ that depends only on M;*

(iii) *(Smallness) f_{sml} has an L^2 norm of at most ε; and*

(iv) *(Pseudorandomness) One has $|\mathbf{E}_{n\in[N]} f_{\mathrm{psd}}(n) e(-\alpha n)| \leq 1/F(M)$ for all $\alpha \in \mathbf{R}$.*

Proof. We recursively define a sequence $M_1 < M_2 < \ldots$ by setting $M_1 := 1$ and $M_{k+1} := M_k + F(M_k) + 1$ (say). Applying Proposition 1.2.10 (starting with the trivial factor $\mathcal{B}_1$), one can then find a nested sequence of refinements $\mathcal{B}_1 \subset \mathcal{B}_2 \subset \ldots$, such that

$$|\langle f - \mathbf{E}(f|\mathcal{B}_k), e(\alpha \cdot)\rangle_{L^2([N])}| < 1/M_k$$

for all $k \geq 1$ and $\alpha \in \mathbf{R}$, and such that each $\mathcal{B}_k$ consists of $O_k(1)$ atoms that are Fourier-measurable with some growth function depending on M_k (note that this quantity dominates k and $M_1, \ldots, M_{k-1}$ by construction). By Pythagoras' theorem, the energies $\|\mathbf{E}(f|\mathcal{B}_k)\|_{L^2([N])}^2$ are monotone increasing between 0 and 1, so by the pigeonhole principle there exists $k = O(1/\varepsilon^2)$ such that

$$\|\mathbf{E}(f|\mathcal{B}_{k+1})\|_{L^2([N])}^2 - \|\mathbf{E}(f|\mathcal{B}_k)\|_{L^2([N])}^2 \leq \varepsilon^2$$

1.2. Roth's theorem

and hence by Pythagoras

$$\|\mathbf{E}(f|\mathcal{B}_{k+1}) - \mathbf{E}(f|\mathcal{B}_k)\|^2_{L^2([N])} \leq \varepsilon^2.$$

Setting $f_{\mathrm{str}} := \mathbf{E}(f|\mathcal{B}_k)$, $f_{\mathrm{sml}} := \mathbf{E}(f|\mathcal{B}_{k+1}) - \mathbf{E}(f|\mathcal{B}_k)$, $f_{\mathrm{psd}} := f - \mathbf{E}(f|\mathcal{B}_{k+1})$, we obtain the claim. □

Remark 1.2.12. This result is essentially due to Green [**Gr2005b**] (though not quite in this language). Earlier related decompositions are due to Bourgain [**Bo1986**] and to Green and Konyagin [**GrKo2009**]. The *Szemerédi regularity lemma in graph theory* can be viewed as the graph-theoretic analogue of this Fourier-analytic result; see [**Ta2006**], [**Ta2007**] for further discussion. The double iteration required to prove Theorem 1.2.11 means that the bounds here are quite poor (of tower-exponential type, in fact, when F is exponential, which is typical in applications), much as in the graph theory case; thus the use of this lemma, while technically quantitative in nature, gives bounds that are usually quite inferior to what is known or suspected to be true.

As with the equidistribution theorems from the previous sections, it is crucial that the uniformity $1/F(M)$ for the pseudorandom component f_{psd} is of an arbitrarily higher quality than the measurability of the structured component f_{str}.

Much as the equidistribution theorems from the previous sections could be used to prove multiple recurrence theorems, the arithmetic regularity lemma can be used (among other things) to give a proof of Roth's theorem. We do so as follows. Let N be a large integer, and let A be a subset of $[N]$ with $|A| \geq \delta N$ for some $\delta > 0$. We consider the expression $\Lambda(1_A, 1_A, 1_A)$, where Λ is the trilinear form

$$\Lambda(f, g, h) := \frac{1}{N^2} \sum_{n \in [N]} \sum_{r \in [-N, N]} f(n) g(n+r) h(n+2r).$$

We will show that

(1.14) $$\Lambda(1_A, 1_A, 1_A) \gg_\delta 1,$$

which implies that the number of all three-term arithmetic progressions in A (including the degenerate ones with $r = 0$) is $\gg_\delta N^2$. For N sufficiently large depending on δ, this number is larger than the number N of degenerate progressions, giving the theorem.

It remains to establish (1.14). We apply Theorem 1.2.11 with parameters $\varepsilon > 0$, F to be chosen later (they will depend on δ) to obtain a quantity M and a decomposition

$$1_A = f_{\mathrm{str}} + f_{\mathrm{sml}} + f_{\mathrm{psd}}$$

with the stated properties. This splits the left-hand side of (1.14) into 27 terms; but we can eliminate several of these terms:

Exercise 1.2.11. Show that all of the terms in (1.14) which involve at least one copy of f_{psd} are of size $O(1/F(M))$. (*Hint:* Modify the proof of Proposition 1.2.4.)

From this exercise we see that

$$(1.15) \quad \Lambda(1_A, 1_A, 1_A) = \Lambda(f_{\text{str}} + f_{\text{sml}}, f_{\text{str}} + f_{\text{sml}}, f_{\text{str}} + f_{\text{sml}}) + O(1/F(M)).$$

Now we need to deal with $f_{\text{str}} + f_{\text{sml}}$. A key point is the *almost periodicity* of $f_{\text{str}} + f_{\text{sml}}$:

Lemma 1.2.13 (Almost periodicity). *For $\gg_{\delta,M} N$ values of $r \in [-\varepsilon N, \varepsilon N]$, one has*

$$\mathbf{E}_{n \in [N]} |(f_{\text{str}} + f_{\text{sml}})(n+r) - (f_{\text{str}} + f_{\text{sml}})(n)| \ll \varepsilon$$

(where we extend $f_{\text{str}}, f_{\text{sml}}$ by zero outside of $[N]$).

Proof. As f_{str} is Fourier-measurable, we can approximate it to an error of $O(\varepsilon)$ in $L^1[N]$ norm by a function

$$(1.16) \quad g = \sum_{j=1}^{J} c_j e(\alpha_j n)$$

of Fourier complexity $J \leq O_{M,\varepsilon}(1)$. From the smallness of f_{sml}, we then have

$$\mathbf{E}_{n \in [N]} |(f_{\text{str}} + f_{\text{sml}})(n+r) - (f_{\text{str}} + f_{\text{sml}})(n)|$$
$$\leq \mathbf{E}_{n \in [N]} |g(n+r) - g(n)| + O(\varepsilon)$$

(where we extend g using (1.16) rather than by zero, with the error being $O(\varepsilon)$ when $|r| \leq \varepsilon N$). We can use (1.16) and the triangle inequality to bound

$$\mathbf{E}_{n \in [N]} |g(n+r) - g(n)| \leq \sum_{j=1}^{J} |e(\alpha_j r) - 1|.$$

Using multiple recurrence, we can find $\gg_{J,\varepsilon} N$ values of $r \in [-\varepsilon N, \varepsilon N]$ such that $\|\alpha_j r\|_{\mathbf{R}/\mathbf{Z}} \leq \varepsilon/J$ for all $1 \leq j \leq J$. The claim follows. $\square$

Now we can finish the proof of Roth's theorem. As $f_{\text{str}} + f_{\text{sml}}$ has the same mean as f, we have

$$\mathbf{E}_{n \in [N]}(f_{\text{str}} + f_{\text{sml}})(n) \geq \delta$$

and hence by Hölder's inequality (and the non-negativity of $f_{\text{str}} + f_{\text{sml}}$),

$$\mathbf{E}_{n \in [N]}(f_{\text{str}} + f_{\text{sml}})(n)^3 \geq \delta^3.$$

Now if r is one of the periods in the above lemma, we have
$$\mathbf{E}_{n\in[N]}|(f_{\text{str}} + f_{\text{sml}})(n+r) - (f_{\text{str}} + f_{\text{sml}})(n)| \ll \varepsilon$$
and thus by shifting
$$\mathbf{E}_{n\in[N]}|(f_{\text{str}} + f_{\text{sml}})(n+2r) - (f_{\text{str}} + f_{\text{sml}})(n+r)| \ll \varepsilon$$
and so by the triangle inequality
$$\mathbf{E}_{n\in[N]}|(f_{\text{str}} + f_{\text{sml}})(n+2r) - (f_{\text{str}} + f_{\text{sml}})(n)| \ll \varepsilon.$$
Putting all this together using the triangle and Hölder inequalities, we obtain
$$\mathbf{E}_{n\in[N]}(f_{\text{str}} + f_{\text{sml}})(n)(f_{\text{str}} + f_{\text{sml}})(n+r)(f_{\text{str}} + f_{\text{sml}})(n+2r) \geq \delta^3 - O(\varepsilon).$$
Thus, if ε is sufficiently small depending on δ, we have
$$\mathbf{E}_{n\in[N]}(f_{\text{str}} + f_{\text{sml}})(n)(f_{\text{str}} + f_{\text{sml}})(n+r)(f_{\text{str}} + f_{\text{sml}})(n+2r) \gg \delta^3$$
for $\gg_{J,\varepsilon} N$ values of r, and thus
$$\Lambda(f_{\text{str}} + f_{\text{sml}}, f_{\text{str}} + f_{\text{sml}}, f_{\text{str}} + f_{\text{sml}}) \gg_{\delta,M} 1;$$
if we then set F to be a sufficiently rapidly growing function (depending on δ), we obtain the claim from (1.15). This concludes the proof of Roth's theorem.

Exercise 1.2.12. Use the energy increment method to establish a different proof of Exercise 1.2.7. (*Hint:* For the multiple recurrence step, use a pigeonhole principle argument rather than an appeal to equidistribution theory.)

We now briefly indicate how to translate the above arguments into the ultralimit setting. We first need to construct an important measure on limit sets, namely *Loeb measure*.

Exercise 1.2.13 (Construction of Loeb measure)**.** Let N be an unbounded natural number. Define the *Loeb measure* $\mu(A)$ of a limit subset A of $[N]$ to be the quantity $\text{st}(|A|/N)$, thus for instance a set of cardinality $o(N)$ will have Loeb measure zero.

(i) Show that if a limit subset A of $[N]$ is partitioned into countably many disjoint limit subsets A_n, that all but finitely many of the A_n are empty, and so $\mu(A) = \mu(A_1) + \cdots + \mu(A_n)$.

(ii) Define the *outer measure* $\mu_*(A)$ of a subset A of $[N]$ (not necessarily a limit subset) to be the infimum of $\sum_n \mu(A_n)$, where $A_1, A_2, \ldots$ is a countable family of limit subsets of $[N]$ that cover A, and call a subset of $[N]$ *null* if it has zero outer measure. Call a subset *Loeb measurable* if it differs from a limit set by a null set. Show that there is a unique extension of Loeb measure μ from limit sets to Loeb measurable sets that is a countably additive probability measure on $[N]$. (*Hint:* Use the *Carathéodory extension theorem*, see e.g. [**Ta2011**, §1.7].)

(iii) If $f: [N] \to \mathbf{C}$ is a limit function bounded in magnitude by some standard real M, show that $\operatorname{st}(f)$ is a Loeb measurable function in $L^\infty(\mu)$, with norm at most M.

(iv) Show that there exists a unique trilinear form $\Lambda: L^\infty(\mu) \times L^\infty(\mu) \times L^\infty(\mu) \to \mathbf{C}$, jointly continuous in the $L^3(\mu)$ topology for all three inputs, such that
$$\Lambda(\operatorname{st}(f), \operatorname{st}(g), \operatorname{st}(h))$$
$$= \operatorname{st}(\frac{1}{N^2} \sum_{n \in [N]} \sum_{r \in [-N,N]} f(n)g(n+r)h(n+2r))$$
for all bounded limit functions f, g, h.

(v) Show that Roth's theorem is equivalent to the assertion that $\Lambda(f, f, f) > 0$ whenever $f \in L^\infty(\mu)$ is a bounded non-negative function with $\int_{[N]} f \, d\mu > 0$.

Loeb measure was introduced in [**Lo1975**], establishing a link between standard and non-standard measure theory.

Next, we develop the ultralimit analogue of Fourier measurability, which we will rename *Kronecker measurability* due to the close analogy with the *Kronecker factor* in ergodic theory.

Exercise 1.2.14 (Construction of the Kronecker factor). Let N be an unbounded natural number. We define a *Fourier character* to be a function in $L^\infty([N])$ of the form $n \mapsto \operatorname{st}(e(\alpha n))$ for some limit real number α. We define a *trigonometric polynomial* to be any finite linear combination (over the standard complex numbers) of Fourier characters. Let $\mathcal{Z}^1$ be the σ-algebra of Loeb measurable sets generated by the Fourier characters; we refer to $\mathcal{Z}^1$ as the *Kronecker factor*, and functions or sets measurable in this factor as *Kronecker measurable* functions and sets. Thus, for instance, all trigonometric polynomials are Kronecker measurable. We let $\mathbf{E}(f|\mathcal{Z}^1)$ denote the orthogonal projection from f to $L^2(\mathcal{Z}^1)$, i.e., the conditional expectation to the Kronecker factor.

(i) Show that if $f \in L^\infty(\mathcal{Z}^1)$ is bounded in magnitude by M and $\varepsilon > 0$ is a standard real, then there exists a trigonometric polynomial $P \in L^\infty(\mathcal{Z}^1)$ which is also bounded in magnitude by M and is within ε of f in L^1 norm.

(ii) Show that if $f \in L^\infty(\mathcal{Z}^1)$ and $\varepsilon > 0$, then there exists a limit subset R of $[-\varepsilon N, \varepsilon N]$ of cardinality $\gg N$ such that $\|f(\cdot) - f(\cdot + r)\|_{L^1([N])} \leq \varepsilon$ for all $r \in R$ (extending f by zero).

(iii) Show that if $f \in L^\infty(\mathcal{Z}^1)$ is non-negative with $\int_{[N]} f \, d\mu > 0$, then $\Lambda(f, f, f) > 0$.

(iv) Show that if $f_1, f_2, f_3 \in L^\infty([N])$ and $\mathbf{E}(f_i|\mathcal{Z}^1) = 0$ for at least one $i = 1, 2, 3$, then $\Lambda(f_1, f_2, f_3) = 0$.

(v) Conclude the proof of Roth's theorem using ultralimits.

Remark 1.2.14. Note how the (finitary) arithmetic regularity lemma has been replaced by the more familiar (infinitary) theory of conditional expectation to a factor, and the finitary notion of measurability has been replaced by a notion from the traditional (countably additive) infinitary theory of measurability. This is one of the key advantages of the ultralimit approach, namely that it allows one to exploit already established theories of infinitary mathematics (e.g. measure theory, ergodic theory, Hilbert space geometry, etc.) to prove a finitary result.

Exercise 1.2.15. Use the ultralimit energy increment method to establish yet another proof of Exercise 1.2.7.

Remark 1.2.15. The ultralimit approach to the above type of decompositions can be generalised to the task of counting more complicated patterns than arithmetic progressions; see [**Sz2009**], [**Sz2009b**], [**Sz2010**]. The approach taken in those papers is analogous in many ways to the ergodic-theoretic approach of Host and Kra [**HoKr2005**], which we will not discuss in detail here.

1.2.3. More quantitative bounds (optional).
The above proofs of Roth's theorem (as formulated in, say, Theorem 1.2.2) were qualitative in the sense that they did not explicitly give a bound for N_0 in terms of δ. Nevertheless, by analysing the finitary arguments more carefully, a bound can be extracted:

Exercise 1.2.16. Show that in Proposition 1.2.6, one can take $N' \gg \varepsilon^{O(1)} N^{1/2}$. Using this and the density increment argument, show that one can take $N_0 \ll \exp(\exp(O(1/\delta)))$ in Theorem 1.2.2. (To put it another way, subsets of $[N]$ of density much larger than $1/\log \log N$ will contain progressions of length three.)

Exercise 1.2.17. Show that in the energy increment proof of Roth's theorem, one can take the growth functions $\mathcal{F}$ involved to be polynomial in K (but with the exponent growing exponentially with each refinement of the factor), and F can be taken to be an iterated exponential; thus ultimately allows one to take N_0 to be a *tower exponential*[6] of height $O(\delta^{-O(1)})$. Thus

---
[6]To put it another way, subsets of $[N]$ of density much larger than $1/\log_*^c N$ for some $c > 0$ will contain progressions of length three, where $\log_* N$ is the number of logarithms needed to reduce N to below (say) 2.

we see that the energy increment argument, in the form presented here, provides much worse bounds than the density increment argument; but see below.

For the ultralimit arguments, it is significantly harder to extract a quantitative bound from the argument (basically one has to painstakingly "finitise" the argument first, essentially reaching the finitary counterparts of these arguments presented above). Thus we see that there is a tradeoff when using ultralimit analysis; the arguments become slightly cleaner (and one can deploy infinitary methods), but one tends to lose sight[7] of what quantitative bounds the method establishes.

It is possible to run the density increment argument more efficiently by combining it with some aspects of the energy increment argument. As described above, the density increment argument proceeds by locating a *single* large Fourier coefficient $\hat{1}_A(\alpha)$ of A, and uses this to obtain a density increment on a relatively short subprogression of $[N]$ (of length comparable to $\sqrt{N}$, ignoring factors of δ). One then has to iterate this about $1/\delta$ times before one obtains a truly significant density increment (e.g. from δ to 2δ). It is this repeated passage from N to $\sqrt{N}$ which is ultimately responsible for the double exponential bound for N_0 at the end of the day.

In an unpublished work, Endre Szemerédi observed that one can run this argument more efficiently by collecting *several* large Fourier coefficients of 1_A simultaneously (somewhat in the spirit of the energy increment argument), and only then passing to a subprogression on which all of the relevant Fourier characters are simultaneously close to constant. The subprogression obtained is smaller as a consequence, but the density increment is much more substantial. Using this strategy, Endre was able to improve the original Roth bound of $N_0 \ll \exp(\exp(O(1/\delta)))$ to the somewhat better $N_0 \ll \exp(\exp(O(\log^2(1/\delta))))$ (or equivalently, he was able to establish length three progressions in any subset of $[N]$ of density much larger than $\exp(-c\sqrt{\log \log N})$ for some $c > 0$). By carefully optimising the choice of threshold for selecting the "large Fourier coefficients", Szemerédi (unpublished) and Heath-Brown [**HB1987**] independently improved this method further to obtain $N_0 \ll \exp(\delta^{-O(1)})$, or equivalently obtaining length three progressions in sets[8] in $[N]$ of density much larger than $\log^{-c} N$.

The next advance was by Bourgain [**Bo1999**], who realised that rather than pass to short subprogressions, it was more efficient to work on the significantly larger (but messier) Bohr sets $\{n : \alpha n \bmod 1 \in I\}$, after ensuring

---

[7]This is particularly the case if one begins to rely heavily on the axiom of choice (or on large cardinal axioms) once one takes ultralimits, although these axioms are not used in the examples above.

[8]This result was later extended to arbitrary finite abelian groups by Meshulam [**Me1995**].

that such Bohr sets were *regular* (this condition is closely related to the Fourier measurability condition used in the energy increment argument). With this modification to the original Roth argument, the bound was lowered to $N_0 \ll \delta^{-O(1/\delta^2)}$, or equivalently obtaining length three progressions in sets of density much larger than $\sqrt{\log \log N / \log N}$. Even more recently, this argument was combined with the Szemerédi-Heath-Brown argument by Bourgain [**Bo2008**], and refined further by Sanders [**Sa2010**], to obtain the further improvement of $N_0 \ll \exp(O(\delta^{-4/3-o(1)}))$, and then (by a somewhat different argument of Sanders [**Sa2010**]) of $N_0 \ll \exp(O(\delta^{-1-o(1)}))$. This is tantalisingly close to the $k=3$ case of an old conjecture of Erdős that asserts that any subset of the natural numbers whose sums of reciprocals diverge should have infinitely many arithmetic progressions of length k for any k. To establish the $k=3$ case from quantitative versions of Roth's theorem, one would basically need a bound of the form $N_0 \ll \exp(\delta^{-1+c})$ for some $c > 0$ (or the ability to obtain progressions in sets of density $1/\log^{1+c} N$). Very recently, a bound of this shape has been obtained in the bounded characteristic case; see [**BaKa2011**].

On the other hand, there is an old counterexample of Behrend [**Be1946**] (based ultimately on the observation that a sphere in a high-dimensional lattice $\mathbf{Z}^d$ does not contain any arithmetic progressions of length three) which shows that N_0 must be at least $\gg \exp(\log^2(1/\delta))$ (in particular, it must be super-polynomial in δ); equivalently, it is known that there are subsets of $[N]$ of density about $\exp(-c\sqrt{\log N})$ with no arithmetic progressions of length three. For the sharpest results in this direction, see [**El2008**] and [**GrWo2008**].

The question of refining the bounds is an important one, as it tends to improve the technological understanding of current methods, as well as shed light on their relative strengths and weaknesses. However, this comes at the cost of making the arguments somewhat more technical, and so we shall not focus on the sharpest quantitative results in this section.

1.3. Linear patterns

In Section 1.2, we used (linear) Fourier analysis to control the number of three-term arithmetic progressions $a, a+r, a+2r$ in a given set A. The power of the Fourier transform for this problem ultimately stemmed from the identity

$$(1.17) \quad \begin{aligned} \mathbf{E}_{n,r \in \mathbf{Z}/N'\mathbf{Z}} 1_A(n) 1_A(n+r) 1_A(n+2r) \\ = \sum_{\alpha \in \frac{1}{N'}\mathbf{Z}/\mathbf{Z}} \hat{1}_A(\alpha) \hat{1}_A(-2\alpha) \hat{1}_A(\alpha) \end{aligned}$$

for any cyclic group $\mathbf{Z}/N'\mathbf{Z}$ and any subset A of that group (analogues of this identity also exist for other finite abelian groups, and to a lesser extent to non-abelian groups also, although that is not the focus of my current discussion).

As it turns out, linear Fourier analysis is not able to discern higher order patterns, such as arithmetic progressions of length four; we give some demonstrations of this below the fold, taking advantage of the polynomial recurrence theory from Section 1.1.

The main objective of this text is to introduce the (still nascent) theory of *higher order Fourier analysis*, which is capable of studying higher order patterns. The full theory is still rather complicated (at least, at our present level of understanding). However, one aspect of the theory is relatively simple, namely that we can largely reduce the study of arbitrary additive patterns to the study of a single type of additive pattern, namely the *parallelopipeds*

(1.18) $$(x + \omega_1 h_1 + \cdots + \omega_d h_d)_{\omega_1,\ldots,\omega_d \in \{0,1\}}.$$

Thus for instance, for $d = 1$ one has the *line segments*

(1.19) $$x, x + h_1;$$

for $d = 2$ one has the *parallelograms*

(1.20) $$x, x + h_1, x + h_2, x + h_1 + h_2;$$

for $d = 3$ one has the *parallelopipeds*
(1.21)
$$x, x+h_1, x+h_2, x+h_3, x+h_1+h_2, x+h_1+h_3, x+h_2+h_3, x+h_1+h_2+h_3.$$

These patterns are particularly pleasant to handle, thanks to the large number of symmetries available on the discrete cube $\{0,1\}^d$. For instance, whereas establishing the presence of arbitrarily long arithmetic progressions in dense sets is quite difficult (cf. Szemerédi's theorem [**Sz1975**]), establishing arbitrarily high-dimensional parallelopipeds is much easier:

Exercise 1.3.1. Let $A \subset [N]$ be such that $|A| > \delta N$ for some $0 < \delta \leq 1$. If N is sufficiently large depending on δ, show that there exists an integer $1 \leq h \ll 1/\delta$ such that $|A \cap (A+h)| \gg \delta^2 N$. (*Hint:* Obtain upper and lower bounds on the set $\{(x,y) \in A \times A : x < y \leq x + 10/\delta\}$.)

Exercise 1.3.2 (Hilbert cube lemma). Let $A \subset [N]$ be such that $|A| > \delta N$ for some $0 < \delta \leq 1$, and let $d \geq 1$ be an integer. Show that if N is sufficiently large depending on δ, d, then A contains a parallelopiped of the form (1.18), with $1 \leq h_1, \ldots, h_d \ll_\delta 1$ positive integers. (*Hint:* Use the previous exercise and induction.) Conclude that if $A \subset \mathbf{Z}$ has positive upper density, then it contains infinitely many such parallelopipeds for each d.

1.3. Linear patterns

Exercise 1.3.3. Show that if $q \geq 1$ is an integer, and d is sufficiently large depending on q, then for any parallelopiped (1.18) in the integers $\mathbf{Z}$, there exists $\omega_1, \ldots, \omega_d \in \{0,1\}$, not all zero, such that $x + h_1\omega_1 + \cdots + h_d\omega_d = x \bmod q$. (*Hint:* Pigeonhole the h_i in the residue classes modulo q.) Use this to conclude that if A is the set of all integers n such that $|n - km!| \geq m$ for all integers $k, m \geq 1$, then A is a set of positive upper density (and also positive lower density) which does not contain any infinite parallelopipeds (thus one cannot take $d = \infty$ in the Hilbert cube lemma).

The standard way to control the parallelogram patterns (and thus, all other (finite complexity) linear patterns) are the *Gowers uniformity norms* (1.22)
$$\|f\|_{U^d(G)} := \mathbf{E}_{x,h_1,\ldots,h_d \in G} \prod_{\omega_1,\ldots,\omega_d \in \{0,1\}^d} \mathcal{C}^{\omega_1 + \cdots + \omega_d} f(x + \omega_1 h_1 + \cdots + \omega_d h_d)$$

with $f: G \to \mathbf{C}$ a function on a finite abelian group G, and $\mathcal{C}: z \mapsto \overline{z}$ is the complex conjugation operator; analogues of this norm also exist for group-like objects such as the progression $[N]$, and also for measure-preserving systems (where they are known as the *Gowers-Host-Kra uniformity semi-norms*, see [**HoKr2005**] for more discussion). In this section we will focus on the basic properties of these norms; the deepest fact about them, known as the *inverse conjecture* for these norms, will be discussed in later sections.

1.3.1. Linear Fourier analysis does not control length four progressions.

Let $A \subset \mathbf{Z}/N\mathbf{Z}$ be a subset of a cyclic group $\mathbf{Z}/N\mathbf{Z}$ with density $|A| = \delta N$; we think of $0 < \delta \leq 1$ as being fixed, and N as being very large or going off to infinity.

For each $k \geq 1$, consider the number

(1.23) $\quad \{(n,r) \in \mathbf{Z}/N\mathbf{Z} \times \mathbf{Z}/N\mathbf{Z} : n, n+r, \ldots, n+(k-1)r \in A\}$

of k-term arithmetic progressions in A (including degenerate progressions). Heuristically, this expression should typically be close to $\delta^k N^2$. Since there are N^2 pairs (n,r) and we would expect each pair to have a δ^k "probability" that $n, n+r, \ldots, n+(k-1)r$ simultaneously lie in A. Indeed, using standard probabilistic tools such as *Chernoff's inequality*, it is not difficult to justify this heuristic with probability asymptotically close to 1 in the case that A is a randomly chosen set of the given density.

Let's see how this heuristic holds up for small values of k. For $k = 0, 1, 2$, this prediction is exactly accurate (with no error term) for any set A with cardinality δN; no randomness hypothesis of any sort is required. For $k = 3$, we see from (1.17) and the observation that $\hat{1}_A(0) = \delta$ that (1.23) is given by the formula

$$N^2 \left(\delta^3 + \sum_{\xi \in \mathbf{Z}/N\mathbf{Z}: \xi \neq 0} \hat{1}_A(\xi)^2 \hat{1}_A(-2\xi) \right).$$

Let us informally say that A is *Fourier-pseudorandom* if one has

$$\sup_{\xi \in \mathbf{Z}/N\mathbf{Z}: \xi \neq 0} |\hat{1}_A(\xi)| = o(1)$$

where $o(1)$ is a quantity that goes to zero as $N \to \infty$. Then from applying Plancherel's formula and Cauchy-Schwarz as in the previous sections, we see that the number of three-term arithmetic progressions is

$$N^2(\delta^3 + o(1)).$$

Thus we see that the Fourier-pseudorandomness hypothesis allows us to count three-term arithmetic progressions almost exactly.

On the other hand, without the Fourier-pseudorandomness hypothesis, the count (1.23) can be significantly different from $\delta^3 N^2$. For instance, if A is an interval $A = [\delta N]$, then it is not hard to see that (1.23) is comparable to $\delta^2 N^2$ rather than $\delta^3 N^2$; the point is that with a set as structured as an interval, once n and $n + r$ lie in A, there is already a very strong chance that $n + 2r$ lies in A also. In the other direction, a construction of Behrend (mentioned in the previous sections) shows the quantity (1.23) can in fact dip below $\delta^C N^2$ for any fixed C (and in fact one can be as small as $\delta^{c \log \frac{1}{\delta}} N^2$ for some absolute constant $c > 0$).

Now we consider the $k = 4$ case of (1.23), which counts four-term progressions. Here, it turns out that Fourier-pseudorandomness is insufficient; it is possible for the quantity (1.23) to be significantly larger or smaller than $\delta^4 N^2$ even if A is pseudorandom, as was observed by Gowers [**Go1998**] (with a closely related observation in the context of ergodic theory by Furstenberg [**Fu1990**]).

Exercise 1.3.4. Let α be an irrational real number, let $0 < \delta < 1$, and let $A := \{n \in [N] : 0 \leq \{\alpha n^2\} \leq \delta\}$. Show that A is Fourier-pseudorandom (keeping α and δ fixed and letting $N \to \infty$). (*Hint:* One can use Exercise 1.1.21 to show that sums of the form $\mathbf{E}_{n \in [N]} e(k\alpha n^2) e(\xi n)$ cannot be large.)

Exercise 1.3.5. Continuing the previous exercise, show that the expression (1.23) for $k = 4$ is equal to $(c\delta^3 + o(1))N^2$ as $N \to \infty$, for some absolute constant $c > 0$, if $\delta > 0$ is sufficiently small. (*Hint:* First show, using the machinery in Section 1.1, that the two-dimensional sequence $(n, r) \mapsto (\alpha n^2, \alpha(n+r)^2, \alpha(n+2r)^2, \alpha(n+3r)^2)$ mod $\mathbf{Z}^4$ is asymptotically equidistributed in the torus $\{(x_1, x_2, x_3, x_4) \in \mathbf{T}^4 : x_1 - 3x_2 + 3x_3 - x_4 = 0\}$.)

1.3. Linear patterns

The above exercises show that a Fourier-pseudorandom set can have a four-term progression count (1.23) significantly larger than $\delta^4 N$. One can also make the count significantly smaller than $\delta^4 N$ (an observation of Gowers, discussed at [**Wo**]), but this requires more work.

Exercise 1.3.6. Let $0 < \delta < 1$. Show that there exists a function $f\colon \mathbf{T}^2 \to [0,1]$ with $\int_{\mathbf{T}} f(x,y)\, dy = \delta$ for all $x \in \mathbf{T}$, such that the expression

(1.24) $$\int_V f(x_1, y_1) \ldots f(x_4, y_4)$$

is strictly less than δ^4, where $V \leq (\mathbf{T}^2)^4$ is the subspace of quadruplets $((x_1, y_1), \ldots, (x_4, y_4))$ such that $x_1, \ldots, x_4$ is in arithmetic progression (i.e., $x_i = x + ir$ for some $x, r \in \mathbf{T}$) and the $y_1, \ldots, y_4$ obey the constraint

$$y_1 - 3y_2 + 3y_3 - y_4 = 0.$$

(*Hint:* Take f of the form
$$f(x,y) := \delta + \varepsilon(f_1(x)\cos(2\pi y) + f_3(x)\cos(6\pi y))$$
where $\varepsilon > 0$ is a small number, and f_1, f_3 are carefully chosen to make the ε^2 term in (1.24) negative.)

Exercise 1.3.7. Show that there exists an absolute constant $c > 0$ such that for all sufficiently small $\delta > 0$ and sufficiently large N (depending on δ) and a set $A \subset [N]$ with $|A| \geq \delta N$, such that (1.23) with $k = 4$ is less than $\delta^{4+c} N^2$. (*Hint:* Take $\delta \sim 2^{-m}$ for some $m \geq 1$, and let A be a random subset of $[N]$ with each element n of $[N]$ lying in A with an independent probability of

$$\prod_{j=1}^{m} f(\alpha_j n \bmod 1, \alpha_j n^2 \bmod 1),$$

where f is the function in the previous exercise (with $\delta = 1/2$), and $\alpha_1, \ldots, \alpha_m$ are real numbers which are linearly independent over $\mathbf{Z}$ modulo 1.)

1.3.2. The 100% case.

Now we consider the question of counting more general linear (or affine) patterns than arithmetic progressions. A reasonably general setting is to count patterns of the form

$$\Psi(\vec{x}) := (\psi_1(\vec{x}), \ldots, \psi_t(\vec{x}))$$

in a subset A of a finite abelian group G (e.g. a cyclic group $G = \mathbf{Z}/N\mathbf{Z}$), where $\vec{x} = (x_1, \ldots, x_d) \in G^d$, and the $\psi_1, \ldots, \psi_t \colon G^d \to G$ are affine-linear forms

$$\psi_i(x_1, \ldots, x_d) = c_i + \sum_{j=1}^{d} c_{i,j} x_j$$

for some fixed integers $c_i, c_{i,j} \in \mathbf{Z}$. To avoid degeneracies, we will assume that all the ψ_i are surjective (or equilently, that the $c_{i,1}, \ldots, c_{i,d}$ do not have a common factor that divides the order of G). This count would then be given by

$$|G|^d \Lambda_\Psi(1_A, \ldots, 1_A)$$

where Λ_Ψ is the d-linear form

$$\Lambda_\Psi(f_1, \ldots, f_d) := \mathbf{E}_{\vec{x} \in G^d} f_1(\psi_1(\vec{x})) \ldots f_t(\psi_t(\vec{x})).$$

For instance, the task of counting arithmetic progressions $n, n+r, \ldots, n+(k-1)r$ corresponds to the case $d=2, t=k$, and $\psi_i(x_1, x_2) := x_1 + (i-1)x_2$.

We have the trivial bound

(1.25) $$|\Lambda_\Psi(f_1, \ldots, f_t)| \leq \|f_1\|_{L^\infty(G)} \ldots \|f_t\|_{L^\infty(G)}$$

where

$$\|f\|_{L^\infty(G)} := \sup_{x \in G} |f(x)|.$$

Remark 1.3.1. One can replace the L^∞ norm on f_i in (1.25) with an L^{p_i} norm for various values of $p_1, \ldots, p_t$. The set of all admissible $p_1, \ldots, p_t$ is described by the *Brascamp-Lieb inequality*; see, for instance, [**BeCaChTa2008**] for further discussion. We will not need these variants of (1.25).

Improving this trivial bound turns out to be a key step in the theory of counting general linear patterns. In particular, it turns out that for any $\varepsilon > 0$, one usually has

$$|\Lambda_\Psi(f_1, \ldots, f_t)| < \varepsilon \|f_1\|_{L^\infty(G)} \ldots \|f_t\|_{L^\infty(G)}$$

except when $f_1, \ldots, f_t$ take a very special form (or at least correlate with functions of a very special form, such as linear or higher order characters).

To reiterate: the key to the subject is to understand the *inverse problem* of characterising those functions $f_1, \ldots, f_d$ for which one has

$$|\Lambda_\Psi(f_1, \ldots, f_t)| \geq \varepsilon \|f_1\|_{L^\infty(G)} \ldots \|f_t\|_{L^\infty(G)}.$$

This problem is of most interest (and the most difficult) in the "1% world" when ε is small (e.g. $\varepsilon = 0.01$), but it is also instructive to consider the simpler cases of the "99% world" when ε is very close to one (e.g. $\varepsilon = 0.99$), or the "100% world" when ε is exactly equal to one. In these model cases one can use additional techniques (error-correction and similar techniques (often of a theoretical computer science flavour) in the 99% world, or exact algebraic manipulation in the 100% world) to understand this expression.

Let us thus begin with analysing the 100% situation. Specifically, we assume that we are given functions $f_1, \ldots, f_t \in L^\infty(G)$ with

$$|\Lambda_\Psi(f_1, \ldots, f_t)| = \|f_1\|_{L^\infty(G)} \ldots \|f_t\|_{L^\infty(G)}$$

1.3. Linear patterns

and wish to classify the functions $f_1, \ldots, f_t$ as best we can. We will normalise all the norms on the right-hand side to be one, thus $|f_i(x)| \leq 1$ for all $x \in G$ and $i = 1, \ldots, t$, and

(1.26) $$|\Lambda_\Psi(f_1, \ldots, f_t)| = 1.$$

By the triangle inequality, we conclude that

$$\Lambda_\Psi(|f_1|, \ldots, |f_t|) \geq 1.$$

On the other hand, we have the crude bound

$$\Lambda_\Psi(|f_1|, \ldots, |f_t|) \leq 1.$$

Thus equality occurs, which (by the surjectivity hypothesis on all the ψ_i) shows that $|f_i(x)| = 1$ for all $x \in G$ and $i = 1, \ldots, t$. Thus we may write $f_i(x) = e(\phi_i(x))$ for some phase functions $\phi_i \colon G \to \mathbf{R}/\mathbf{Z}$. We then have

$$\Lambda_\Psi(f_1, \ldots, f_t) = \mathbf{E}_{\vec{x} \in G^d} e\left(\sum_{i=1}^{t} \phi_i(\psi_i(\vec{x}))\right)$$

and so from (1.26) one has the equation

(1.27) $$\sum_{i=1}^{t} \phi_i(\psi_i(\vec{x})) = c$$

for all $\vec{x} \in G^d$ and some constant c.

So the problem now reduces to the algebraic problem of solving functional equations such as (1.27). To illustrate this type of problem, let us consider a simple case when $d = 2, t = 3$ and

$$\psi_1(x, y) = x;\ \psi_2(x, y) = y;\ \psi_3(x, y) = x + y$$

in which case we are trying to understand solutions $\phi_1, \phi_2, \phi_3 \colon G \to \mathbf{R}/\mathbf{Z}$ to the functional equation

(1.28) $$\phi_1(x) + \phi_2(y) + \phi_3(x + y) = c.$$

This equation involves three unknown functions ϕ_1, ϕ_2, ϕ_3. But we can eliminate two of the functions by taking *discrete derivatives*. To motivate this idea, let us temporarily assume that G is the real line $\mathbf{R}$ rather than a finite group, and that the functions ϕ_1, ϕ_2, ϕ_3 are smooth. If we then apply the partial derivative operator ∂_x to the above functional equation, one eliminates ϕ_2 and obtains

$$\phi_1'(x) + \phi_3'(x + y) = 0;$$

applying ∂_y then eliminates ϕ_1 and leaves us with

$$\phi_3''(x + y) = 0,$$

thus ϕ_3'' vanishes identically; we can integrate this twice to conclude that ϕ_3 is a linear function of its input,

$$\phi_3(x) = a_3 x + b_3$$

for some constants $a_3, b_3 \in \mathbf{R}$. A similar argument (using the partial derivative operator $\partial_x - \partial_y$ to eliminate ϕ_3, or by applying change of variables such as $(x, z) := (x, x + y)$) shows that $\phi_1(x) = a_1 x + b_1$ and $\phi_2(x) = a_2 x + b_2$ for some additional constants a_1, b_1, a_2, b_2. Finally, by returning to (1.28) and comparing coefficients we obtain the additional compatibility condition $a_3 = -a_1 = -a_2$, which one then easily verifies to completely describe all possible solutions to this equation in the case of smooth functions on $\mathbf{R}$.

Returning now to the discrete world, we mimic the continuous operation of a partial derivative by introducing difference operators

$$\partial_h \phi(x) := \phi(x + h) - \phi(x)$$

for $h \in G$. If we take differences in (1.28) with respect to the x variable by an arbitrary shift $h \in G$ by replacing x by $x + h$ and then subtracting, we eliminate ϕ_2 and obtain

$$(\partial_h \phi_1)(x) + (\partial_h \phi_3)(x + y) = 0;$$

if we then takes differences with respect to the y variable by a second arbitrary shift $k \in G$, one obtains

$$(\partial_k \partial_h \phi_3)(x + y) = 0$$

for all $x, y, h, k \in G$; in particular, $\partial_k \partial_h \phi_3 \equiv 0$ for all $k, h \in G$. Such functions are affine-linear:

Exercise 1.3.8. Let $\phi: G \to \mathbf{R}/\mathbf{Z}$ be a function. Show that $\partial_k \partial_h \phi = 0$ if and only if one has $\phi(x) = a(x) + b$ for some $b \in G$ and some homomorphism $a: G \to \mathbf{R}/\mathbf{Z}$. Conclude that the solutions to (1.28) are given by the form $\phi_i(x) = a_i(x) + b_i$, where $b_1, b_2, b_3 \in G$ and $a_1, a_2, a_3: G \to \mathbf{R}/\mathbf{Z}$ are homomorphisms with $a_1 = -a_2 = -a_3$.

Having solved the functional equation (1.28), let us now look at an equation related to four term arithmetic progressions, namely

(1.29) $\quad \phi_1(x) + \phi_2(x + y) + \phi_3(x + 2y) + \phi_4(x + 3y) = c$

for all $x, y \in G$, some constant $c \in G$, and some functions $\phi_1, \phi_2, \phi_3, \phi_4: G \to \mathbf{R}/\mathbf{Z}$. We will try to isolate ϕ_4 by using discrete derivatives as before to eliminate the other functions. First, we differentiate in the y direction by an arbitrary shift $h \in G$, leading to

$$(\partial_h \phi_2)(x + y) + (\partial_{2h} \phi_3)(x + 2y) + (\partial_{3h} \phi_4)(x + 3y) = 0.$$

1.3. Linear patterns

In preparation for then eliminating ϕ_2, we shift x backwards by y, obtaining
$$(\partial_h \phi_2)(x) + (\partial_{2h}\phi_3)(x+y) + (\partial_{3h}\phi_4)(x+2y) = 0.$$
Differentiating in the y direction by another arbitrary shift $k \in G$, we obtain
$$(\partial_k \partial_{2h}\phi_3)(x+y) + (\partial_{2k}\partial_{3h}\phi_4)(x+2y) = 0.$$
We shift x backwards by y again:
$$(\partial_k \partial_{2h}\phi_3)(x) + (\partial_{2k}\partial_{3h}\phi_4)(x+y) = 0.$$
One final differentiation in y by an arbitrary shift $l \in G$ gives
$$(\partial_l \partial_{2k}\partial_{3h}\phi_4)(x+y) = 0.$$
For simplicity, we now make the assumption that the order $|G|$ of G is not divisible by either 2 or 3, so that the homomorphisms $k \mapsto 2k$ and $h \mapsto 3h$ are automorphisms of G. We conclude that

(1.30) $$\partial_l \partial_k \partial_h \phi_4 \equiv 0$$

for all l, k, h. Such functions will be called *quadratic* functions from G to $\mathbf{R}/\mathbf{Z}$, thus ϕ_4 is quadratic. A similar argument shows that ϕ_1, ϕ_2, ϕ_3 are quadratic.

Just as (affine-)linear functions can be completely described in terms of homomorphisms, quadratic functions can be described in terms of bilinear forms, as long as one avoids the characteristic 2 case:

Exercise 1.3.9. Let G be a finite abelian group with $|G|$ not divisible by 2. Show that a map $\phi\colon G \to \mathbf{R}/\mathbf{Z}$ is quadratic if and only one has a representation of the form
$$\phi(x) = B(x,x) + L(x) + c$$
where $c \in \mathbf{R}/\mathbf{Z}$, $L\colon G \to \mathbf{R}/\mathbf{Z}$ is a homomorphism, and $B\colon G \times G \to \mathbf{R}/\mathbf{Z}$ is a symmetric bihomomorphism (i.e., $B(x,y) = B(y,x)$, and B is a homomorphism in each of x, y individually (holding the other variable fixed)). (*Hint:* Heuristically, one should set $B(h,k) := \frac{1}{2}\partial_h \partial_k \phi(x)$, but there is a difficulty because the operation of dividing by $\frac{1}{2}$ is not well defined on $\mathbf{R}/\mathbf{Z}$. It is, however, well defined on $|G|^{\text{th}}$ roots of unity, thanks to $|G|$ not being divisible by two. Once B has been constructed, subtract it off and use Exercise 1.3.8.) What goes wrong when $|G|$ is divisible by 2?

Exercise 1.3.10. Show that when $|G|$ is not divisible by $2, 3$, that the complete solution to (1.29) is given by
$$\phi_i(x) = B_i(x,x) + L_i(x) + c_i$$
for $i = 1, 2, 3, 4$, $c_i \in \mathbf{R}/\mathbf{Z}$, homomorphisms $L_i\colon G \to \mathbf{R}/\mathbf{Z}$, and symmetric bihomomorphisms $B_i\colon G \times G \to \mathbf{R}/\mathbf{Z}$ with $B_2 = -3B_1, B_3 = 3B_1, B_4 = -B_1$ and $L_1 + L_2 + L_3 + L_4 = L_2 + 2L_3 + 3L_4 = 0$.

Exercise 1.3.11. Obtain a complete solution to the functional equation (1.29) in the case when $|G|$ is allowed to be divisible by 2 or 3. (This is an open-ended and surprisingly tricky exercise; it of course depends on what one is willing to call a "solution" to the problem. Use your own judgement.)

Exercise 1.3.12. Call a map $\phi\colon G \to \mathbf{R}/\mathbf{Z}$ *a polynomial of degree $\leq d$* if one has $\partial_{h_1} \ldots \partial_{h_{d+1}} \phi(x) = 0$ for all $x, h_1, \ldots, h_{d+1} \in G$. Show that if $k \geq 1$ and $\phi_1, \ldots, \phi_k$ obey the functional equation

$$\phi_1(x) + \phi_2(x+y) + \cdots + \phi_k(x + (k-1)y) = c$$

and $|G|$ is not divisible by any integer between 2 and $k-1$, then $\phi_1, \ldots, \phi_k$ are polynomials of degree $\leq k - 2$.

We are now ready to turn to the general case of solving equations of the form (1.27). We relied on two main tricks to solve these equations: differentiation, and change of variables. When solving an equation such as (1.29), we alternated these two tricks in turn. To handle the general case, it is more convenient to rearrange the argument by doing all the change of variables in advance. For instance, another way to solve (1.29) is to first make the (non-injective) change of variables

$$(x, y) := (b + 2c + 3d, -a - b - c - d)$$

for arbitrary $a, b, c, d \in G$, so that

$$(x, x+y, x+2y, x+3y) = (b+2c+3d, -a+c+2d, -2a-b+d, -3a-2b-c)$$

and (1.29) becomes
(1.31)
$$\phi_1(b+2c+3d) + \phi_2(-a+c+2d) + \phi_3(-2a-b+d) + \phi_4(-3a-2b-c) = \text{const}$$

for all $a, b, c, d \in G$. The point of performing this change of variables is that while the ϕ_4 term (for instance) involves all three variables, a, b, c, the remaining terms only depend on two of the a, b, c variables at a time. If we now pick $h, k, l \in G$ arbitrarily, and then differentiate in the a, b, c variables by the shifts h, k, l, respectively, then we eliminate the ϕ_1, ϕ_2, ϕ_3 terms and arrive at

$$(\partial_{-l}\partial_{-2k}\partial_{-3h}\phi_4)(-3a - 2b - c) = 0$$

which soon places us back at (1.30) (assuming as before that $|G|$ is not divisible by 2 or 3).

Now we can do the general case, once we put in place a definition (from [GrTa2010]):

Definition 1.3.2 (Cauchy-Schwarz complexity). A system $\psi_1, \ldots, \psi_t\colon G^d \to G$ of affine-linear forms (with linear coefficients in $\mathbf{Z}$) have *Cauchy-Schwarz complexity at most s* if, for every $1 \leq i \leq t$, one can partition $[t]\setminus\{i\}$ into $s + 1$ classes (some of which may be empty), such that ψ_i does not lie in

1.3. Linear patterns

the affine-linear span (over $\mathbf{Q}$) of the forms in any of these classes. The *Cauchy-Schwarz complexity* of a system is defined to be the least such s with this property, or ∞ if no such s exists.

The adjective "Cauchy-Schwarz" (introduced by Gowers and Wolf [GoWo2010]) may be puzzling at present, but will be motivated later.

This is a somewhat strange definition to come to grips with at first, so we illustrate it with some examples. The system of forms $x, y, x+y$ is of complexity 1; given any form here, such as y, one can partition the remaining forms into two classes, namely $\{x\}$ and $\{x+y\}$, such that y is not in the affine-linear span of either. On the other hand, as y is in the affine linear span of $\{x, x+y\}$, the Cauchy-Schwarz complexity is not zero.

Exercise 1.3.13. Show that for any $k \geq 2$, the system of forms $x, x+y, \ldots, x+(k-1)y$ has complexity $k-2$.

Exercise 1.3.14. Show that a system of non-constant forms has finite Cauchy-Schwarz complexity if and only if no form is an affine-linear combination of another.

There is an equivalent way to formulate the notion of Cauchy-Schwarz complexity, in the spirit of the change of variables mentioned earlier. Define the *characteristic* of a finite abelian group G to be the least order of a non-identity element.

Proposition 1.3.3 (Equivalent formulation of Cauchy-Schwarz complexity). *Let $\psi_1, \ldots, \psi_t \colon G^d \to G$ be a system of affine-linear forms. Suppose that the characteristic of G is sufficiently large depending on the coefficients of $\psi_1, \ldots, \psi_t$. Then $\psi_1, \ldots, \psi_t$ has Cauchy-Schwarz complexity at most s if and only if, for each $1 \leq i \leq t$, one can find a linear change of variables $\vec{x} = L_i(y_1, \ldots, y_{s+1}, z_1, \ldots, z_m)$ over $\mathbf{Q}$ such that the form $\dot{\psi}_i(L_i(y_1, \ldots, y_{s+1}, z_1, \ldots, z_m))$ has non-zero $y_1, \ldots, y_{s+1}$ coefficients, but all the other forms $\dot{\psi}_j(L_i(y_1, \ldots, y_{s+1}, z_1, \ldots, z_m))$ with $j \neq i$ have at least one vanishing $y_1, \ldots, y_{s+1}$ coefficient, and $\dot{\psi}_i \colon \mathbf{Q}^d \to \mathbf{Q}$ is the linear form induced by the integer coefficients of ψ_i.*

Proof. To show the "only if" part, observe that if $1 \leq i \leq t$ and L_i is as above, then we can partition the ψ_j, $j \neq i$ into $s+1$ classes depending on which y_k coefficient vanishes for $k = 1, \ldots, s+1$ (breaking ties arbitrarily), and then ψ_i is not representable as an affine-linear combination of the forms from any of these classes (here we use the large characteristic hypothesis). Conversely, suppose $\psi_1, \ldots, \psi_t$ has Cauchy-Schwarz complexity at most s, and let $1 \leq i \leq s$. We can then partition the $j \neq i$ into $s+1$ classes $\mathcal{A}_1, \ldots, \mathcal{A}_{s+1}$, such that ψ_i cannot be expressed as an affine-linear combination of the ψ_j from $\mathcal{A}_k$ for any $1 \leq k \leq s+1$. By duality, one can then find vectors $v_k \in \mathbf{Q}^d$ for each $1 \leq k \leq s+1$ such that $\dot\psi_i$ does not annihilate v_k, but all the $\dot\psi_j$ from $\mathcal{A}_k$ do. If we then set

$$L_i(y_1, \ldots, y_{s+1}, z_1, \ldots, z_d) := (z_1, \ldots, z_d) + y_1 v_1 + \cdots + y_{s+1} v_{s+1},$$

then we obtain the claim. $\square$

Exercise 1.3.15. Let $\psi_1, \ldots, \psi_t \colon G^d \to G$ be a system of affine-linear forms with Cauchy-Schwarz complexity at most s, and suppose that the equation (1.27) holds for some finite abelian group G and some $\phi_1, \ldots, \phi_t \colon G \to \mathbf{R}/\mathbf{Z}$. Suppose also that the characteristic of G is sufficiently large depending on the coefficients of $\psi_1, \ldots, \psi_t$. Conclude that all of the $\phi_1, \ldots, \phi_t$ are polynomials of degree $\leq t$.

It turns out that this result is not quite best possible. Define the *true complexity* of a system of affine-linear forms $\psi_1, \ldots, \psi_t \colon G^d \to G$ to be the largest s such that the powers $\dot\psi_1^s, \ldots, \dot\psi_t^s \colon \mathbf{Q}^d \to \mathbf{Q}$ are linearly independent over $\mathbf{Q}$.

Exercise 1.3.16. Show that the true complexity is always less than or equal to the Cauchy-Schwarz complexity, and give an example to show that strict inequality can occur. Also, show that the true complexity is finite if and only if the Cauchy-Schwarz complexity is finite.

Exercise 1.3.17. Show that Exercise 1.3.15 continues to hold if Cauchy-Schwarz complexity is replaced by true complexity. (*Hint:* First understand the cyclic case $G = \mathbf{Z}/N\mathbf{Z}$, and use Exercise 1.3.15 to reduce to the case when all the ϕ_i are polynomials of bounded degree. The main point is to use a "Lefschetz principle" to lift statements in $\mathbf{Z}/N\mathbf{Z}$ to a characteristic zero field such as $\mathbf{Q}$.) Show that the true complexity cannot be replaced by any smaller quantity.

See [GoWo2010] for further discussion of the relationship between Cauchy-Schwarz complexity and true complexity.

1.3.3. The Gowers uniformity norms. In the previous section, we saw that equality in the trivial inequality (1.25) only occurred when the functions $f_1, \ldots, f_t$ were of the form $f_i = e(\phi_i)$ for some polynomials ϕ_i of degree at most s, where s was the true complexity (or Cauchy-Schwarz complexity) of the system $\psi_1, \ldots, \psi_t$. Another way of phrasing this latter fact is that one has the identity

$$\Delta_{h_1} \ldots \Delta_{h_{s+1}} f_i(x) = 1$$

for all $h_1, \ldots, h_{s+1}, x \in G$, where Δ_h is the multiplicative derivative

$$\Delta_h f(x) := f(x+h)\overline{f(x)}.$$

1.3. Linear patterns

This phenomenon extends beyond the "100% world" of exact equalities. For any $f\colon G \to \mathbf{C}$ and $d \geq 1$, we define the *Gowers uniformity norm* $\|f\|_{U^d(G)}$ by the formula

$$\|f\|_{U^d(G)} := (\mathbf{E}_{h_1,\ldots,h_d,x\in G} \Delta_{h_1} \ldots \Delta_{h_d} f(x))^{1/2^d}; \tag{1.32}$$

note that this is equivalent to (1.22). Using the identity

$$\mathbf{E}_{h,x\in G} \Delta_h f(x) = |\mathbf{E}_{x\in G} f(x)|^2$$

we easily verify that the expectation in the definition of (1.32) is a non-negative real. We also have the recursive formula

$$\|f\|_{U^d(G)} := (\mathbf{E}_{h\in G} \|\Delta_h f\|_{U^{d-1}(G)}^{2^{d-1}})^{1/2^d} \tag{1.33}$$

for all $d \geq 1$.

The U^1 norm is essentially just the mean:

$$\|f\|_{U^1(G)} = |\mathbf{E}_{x\in G} f(x)|. \tag{1.34}$$

As such, it is actually a *semi-norm* rather than a norm.

The U^2 norm can be computed in terms of the Fourier transform:

Exercise 1.3.18 (Fourier representation of U^2). Define the *Pontryagin dual* $\hat{G}$ of a finite abelian group G to be the space of all homomorphisms $\xi\colon G \to \mathbf{R}/\mathbf{Z}$. For each function $f\colon G \to \mathbf{C}$, define the Fourier transform $\hat{f}\colon \hat{G} \to \mathbf{C}$ by the formula $\hat{f}(\xi) := \mathbf{E}_{x\in G} f(x) e(-\xi(x))$. Establish the identity

$$\|f\|_{U^2(G)} = \|\hat{f}\|_{\ell^4(\hat{G})} := (\sum_{\xi\in\hat{G}} |\hat{f}(\xi)|^4)^{1/4}.$$

In particular, the U^2 norm is a genuine norm (thanks to the norm properties of $\ell^4(G)$, and the injectivity of the Fourier transform).

For the higher Gowers norms, there is not nearly as nice a formula known in terms of things like the Fourier transform, and it is not immediately obvious that these are indeed norms. But this can be established by introducing the more general *Gowers inner product*

$$\langle (f_\omega)_{\omega\in\{0,1\}^d} \rangle_{U^d(G)} := \mathbf{E}_{x,h_1,\ldots,h_d\in G} \prod_{\omega_1,\ldots,\omega_d\in\{0,1\}^d} \mathcal{C}^{\omega_1+\cdots+\omega_d} f_{\omega_1,\ldots,\omega_d}(x+\omega_1 h_1 + \cdots + \omega_d h_d)$$

for any 2^d-tuple $(f_\omega)_{\omega\in\{0,1\}^d}$ of functions $f_\omega\colon G \to \mathbf{C}$, thus, in particular,

$$\langle (f)_{\omega\in\{0,1\}^d} \rangle_{U^d(G)} = \|f\|_{U^d(G)}^{2^d}.$$

The relationship between the Gowers inner product and the Gowers uniformity norm is analogous to that between a Hilbert space inner product and the Hilbert space norm. In particular, we have the following analogue of the Cauchy-Schwarz inequality:

Exercise 1.3.19 (Cauchy-Schwarz-Gowers inequality). For any tuple $(f_\omega)_{\omega \in \{0,1\}^d}$ of functions $f_\omega \colon G \to \mathbf{C}$, use the Cauchy-Schwarz inequality to show that

$$|\langle (f_\omega)_{\omega \in \{0,1\}^d} \rangle_{U^d(G)}| \leq \prod_{j=0,1} |\langle (f_{\pi_{i,j}(\omega)})_{\omega \in \{0,1\}^d} \rangle_{U^d(G)}|^{1/2}$$

for all $1 \leq i \leq d$, where for $j = 0, 1$ and $\omega \in \{0,1\}^d$, $\pi_{i,j}(\omega) \in \{0,1\}^d$ is formed from ω by replacing the ith coordinate with j. Iterate this to conclude that

$$|\langle (f_\omega)_{\omega \in \{0,1\}^d} \rangle_{U^d(G)}| \leq \prod_{\omega \in \{0,1\}^d} \|f_\omega\|_{U^d(G)}.$$

Then use this to conclude the monotonicity formula

$$\|f\|_{U^d(G)} \leq \|f\|_{U^{d+1}(G)}$$

for all $d \geq 1$, and the triangle inequality

$$\|f + g\|_{U^d(G)} \leq \|f\|_{U^d(G)} + \|g\|_{U^d(G)}$$

for all $f, g \colon G \to \mathbf{C}$. (*Hint:* For the latter inequality, raise both sides to the power 2^d and expand the left-hand side.) Conclude, in particular, that the $U^d(G)$ norms are indeed norms for all $d \geq 2$.

The Gowers uniformity norms can be viewed as a quantitative measure of how well a given function behaves like a polynomial. One piece of evidence in this direction is:

Exercise 1.3.20 (Inverse conjecture for the Gowers norm, 100% case). Let $f \colon G \to \mathbf{C}$ be such that $\|f\|_{L^\infty(G)} = 1$, and let $s \geq 0$. Show that $\|f\|_{U^{s+1}(G)} \leq 1$, with equality if and only if $f = e(\phi)$ for some polynomial $\phi \colon G \to \mathbf{R}/\mathbf{Z}$ of degree at most s.

The problem of classifying smaller values of $\|f\|_{U^{s+1}(G)}$ is significantly more difficult, and will be discussed in later sections.

Exercise 1.3.21 (Polynomial phase invariance). If $f \colon G \to \mathbf{C}$ is a function and $\phi \colon G \to \mathbf{R}/\mathbf{Z}$ is a polynomial of degree at most s, show that $\|e(\phi)f\|_{U^{s+1}(G)} = \|f\|_{U^{s+1}(G)}$. Conclude, in particular, that

$$\sup_\phi |\mathbf{E}_{x \in G} e(\phi(x)) f(x)| \leq \|f\|_{U^{s+1}(G)}$$

where ϕ ranges over polynomials of degree at most s.

1.4. Equidistribution in finite fields

The main utility for the Gowers norms in this subject comes from the fact that they control many other expressions of interest. Here is a basic example:

Exercise 1.3.22. Let $f\colon G \to \mathbf{C}$ be a function, and for each $1 \leq i \leq s+1$, let $g_i\colon G^{s+1} \to \mathbf{C}$ be a function bounded in magnitude by 1 which is independent of the ith coordinate of G^{s+1}. Let $a_1,\ldots,a_{s+1}$ be non-zero integers, and suppose that the characteristic of G exceeds the magnitude of any of the a_i. Show that

$$|\mathbf{E}_{x_1,\ldots,x_{s+1} \in G} f(a_1 x_1 + \cdots + a_{s+1} x_{s+1}) \prod_{i=1}^{s+1} g_i(x_1,\ldots,x_{s+1})|$$

$$\leq \|f\|_{U^{s+1}(G)}.$$

(*Hint:* Induct on s and use (1.33) and the Cauchy-Schwarz inequality.)

This gives us an analogue of Exercise 1.3.15:

Exercise 1.3.23 (Generalised von Neumann inequality). Let $\Psi = (\psi_1,\ldots,\psi_t)$ be a collection of affine-linear forms $\psi_i\colon G^d \to G$ with Cauchy-Schwarz complexity s. If the characteristic of G is sufficiently large depending on the linear coefficients of $\psi_1,\ldots,\psi_t$, show that one has the bound

$$|\Lambda_\Psi(f_1,\ldots,f_t)| \leq \inf_{1 \leq i \leq t} \|f_i\|_{U^{s+1}(G)}$$

whenever $f_1,\ldots,f_t\colon G \to \mathbf{C}$ are bounded in magnitude by one.

Conclude, in particular, that if A is a subset of G with $|A| = \delta|G|$, then

$$\Lambda_\Psi(1_A,\ldots,1_A) = \delta^t + O_t(\|1_A - \delta\|_{U^{s+1}(G)}).$$

From the above inequality, we see that if A has some positive density $\delta > 0$ but has much fewer than $\delta^t N^d/2$ (say) patterns of the form $\psi_1(\vec{x}),\ldots,\psi_t(\vec{x})$ with $\vec{x} \in G^d$, then we have

$$\|1_A - \delta\|_{U^{s+1}(G)} \gg_{t,\delta} 1.$$

This is the initial motivation for studying *inverse theorems* for the Gowers norms, which give necessary conditions for a (bounded) function to have large $U^{s+1}(G)$ norm. This will be a focus of subsequent sections.

1.4. Equidistribution of polynomials over finite fields

In the previous sections, we have focused mostly on the equidistribution or linear patterns on a subset of the integers $\mathbf{Z}$, and in particular on intervals $[N]$. The integers are of course a very important domain to study in additive combinatorics; but there are also other fundamental model examples of domains to study. One of these is that of a vector space V over a finite field $\mathbf{F} = \mathbf{F}_p$ of prime order. Such domains are of interest in computer science (particularly when $p = 2$) and also in number theory; but they also serve as an important simplified *dyadic model* for the integers. See [**Ta2008**, §1.6] or [**Gr2005a**] for further discussion of this point.

The additive combinatorics of the integers $\mathbf{Z}$, and of vector spaces V over finite fields, are analogous, but not quite identical. For instance, the analogue of an arithmetic progression in $\mathbf{Z}$ is a subspace of V. In many cases, the finite field theory is a little bit simpler than the integer theory; for instance, subspaces are closed under addition, whereas arithmetic progressions are only "almost" closed[9] under addition in various senses. However, there are some ways in which the integers are better behaved. For instance, because the integers can be generated by a single generator, a homomorphism from $\mathbf{Z}$ to some other group G can be described by a single group element $g \colon n \mapsto g^n$. However, to specify a homomorphism from a vector space V to G one would need to specify one group element for each dimension of V. Thus we see that there is a tradeoff when passing from $\mathbf{Z}$ (or $[N]$) to a vector space model; one gains a bounded torsion property, at the expense[10] of conceding the bounded generation property.

The starting point for this text (Section 1.1) was the study of equidistribution of polynomials $P \colon \mathbf{Z} \to \mathbf{R}/\mathbf{Z}$ from the integers to the unit circle. We now turn to the parallel theory of equidistribution of polynomials $P \colon V \to \mathbf{R}/\mathbf{Z}$ from vector spaces over finite fields to the unit circle. Actually, for simplicity we will mostly focus on the *classical* case, when the polynomials in fact take values in the pth roots of unity (where p is the characteristic of the field $\mathbf{F} = \mathbf{F}_p$). As it turns out, the non-classical case is also of importance (particularly in low characteristic), but the theory is more difficult; see [**Ta2009**, §1.12] for some further discussion.

1.4.1. Polynomials: basic theory. Throughout this section, V will be a finite-dimensional vector space over a finite field $\mathbf{F} = \mathbf{F}_p$ of prime order p.

Recall from Section 1.3 that a function $P \colon V \to \mathbf{R}/\mathbf{Z}$ is a function is a *polynomial of degree at most d* if

$$\partial_{h_1} \ldots \partial_{h_{d+1}} P(x) = 0$$

for all $x, h_1, \ldots, h_{d+1} \in V$, where $\partial_h P(x) := P(x+h) - P(x)$. As mentioned in previous sections, this is equivalent to the assertion that the Gowers uniformity norm $\|e(P)\|_{U^{d+1}(V)} = 1$. The space of polynomials of degree at most d will be denoted $\text{Poly}_{\leq d}(V \to \mathbf{R}/\mathbf{Z})$; it is clearly an additive

---

[9] For instance, $[N]$ is closed under addition approximately half of the time.

[10] Of course, if one wants to deal with arbitrarily large domains, one has to concede one or the other; the only additive groups that have both bounded torsion and boundedly many generators, are bounded.

1.4. Equidistribution in finite fields

group. Note that a polynomial of degree zero is the same thing as a constant function, thus $\text{Poly}_{\leq 0}(V \to \mathbf{R}/\mathbf{Z}) \equiv \mathbf{R}/\mathbf{Z}$.

An important special case of polynomials are the *classical polynomials*, which take values in $\mathbf{F}$ (which we identify with the pth roots of unity in $\mathbf{R}/\mathbf{Z}$ in the obvious manner); the space of such polynomials of degree at most d will be denoted $\text{Poly}_{\leq d}(V \to \mathbf{F})$; this is clearly a vector space over $\mathbf{F}$. The classical polynomials have a familiar description, once we use a basis to identify V with $\mathbf{F}^n$:

Exercise 1.4.1. Let $P\colon \mathbf{F}^n \to \mathbf{F}$ be a function, and $d \geq 0$ an integer. Show that P is a (classical) polynomial of degree at most d if and only if one has a representation of the form

$$P(x_1, \ldots, x_n) := \sum_{i_1, \ldots, i_n \geq 0 : i_1 + \cdots + i_n \leq d} c_{i_1, \ldots, i_n} x_1^{i_1} \cdots x_n^{i_n}$$

for some coefficients $c_{i_1, \ldots, i_n} \in \mathbf{F}$. Furthermore, show that we can restrict the exponents $i_1, \ldots, i_n$ to lie in the range $\{0, \ldots, p-1\}$, and that once one does so, the representation is unique. (*Hint:* First establish the $d = 1$ case, which can be done, for instance, by a dimension counting argument, and then induct on dimension.)

Exercise 1.4.2. Show that the cardinality of $\text{Poly}_{\leq d}(V \to \mathbf{F})$ is at most $p^{\binom{d+\dim(V)}{d}}$, with equality if and only if $d < p$.

Now we study more general polynomials. A basic fact here is that multiplying a polynomial by the characteristic p lowers the degree:

Lemma 1.4.1. *If* $P \in \text{Poly}_{\leq d}(V \to \mathbf{R}/\mathbf{Z})$, *then* $pP \in \text{Poly}_{\leq \max(d-p+1,0)}(V \to \mathbf{R}/\mathbf{Z})$.

Proof. Without loss of generality, we may take $d \geq p-1$; an easy induction on d then shows it suffices to verify the base case $d = p-1$. Our task is now to show that pP is constant, or equivalently that $p\Delta_e P = 0$ for all $e \in V$.

Fix e. The operator $1 + \Delta_e$ represents a shift by e. Since $pe = 0$, we conclude that $(1 + \Delta_e)^p P = P$. On the other hand, as P has degree at most $p-1$, $\Delta_e^p P = 0$, and so

$$((1+\Delta_e)^p - 1 - \Delta_e^p)P = 0.$$

Using the binomial formula, we can factorise the left-hand side as

$$(1 + \frac{p-1}{2}\Delta_e + \cdots + \Delta_e^{p-2})(p\Delta_e P) = 0.$$

The first factor can be inverted by Neumann series since Δ_e acts nilpotently on polynomials. We conclude that $p\Delta_e P = 0$ as required. $\square$

Exercise 1.4.3. Establish the identity[11]
$$p(T^j - 1) = (-1)^{p-1}(T^j - 1)(T - 1)(T^2 - 1)\ldots(T^{p-1} - 1)$$
$$\mod T^p - 1$$
for an indeterminate T and any integer j, by testing on p^{th} roots of unity. Use this to give an alternate proof of Lemma 1.4.1.

This classifies all polynomials in the high characteristic case $p > d$:

Corollary 1.4.2. *If $p > d$, then* $\text{Poly}_{\leq d}(V \to \mathbf{R}/\mathbf{Z}) = \text{Poly}_{\leq d}(V \to \mathbf{F}) + (\mathbf{R}/\mathbf{Z})$. *In other words, every polynomial of degree at most d is the sum of a classical polynomial and a constant.*

The situation is more complicated in the low characteristic case $p \leq d$, in which non-classical polynomials can occur (polynomials that are not simply a classical polynomial up to constants). For instance, consider the function $P\colon \mathbf{F}_2 \to \mathbf{R}/\mathbf{Z}$ defined by $P(0) = 0$ and $P(1) = 1/4$. One easily verifies that this is a (non-classical) quadratic (i.e., a polynomial of degree at most 2), but is clearly not a shifted version of a classical polynomial since its range is not a shift of the second roots $\{0, 1/2\}$ mod 1 of unity.

Exercise 1.4.4. Let $P\colon \mathbf{F}_2 \to \mathbf{R}/\mathbf{Z}$ be a function. Show that P is a polynomial of degree at most d if and only if the range of P is a translate of the $(2^d)^{\text{th}}$ roots of unity (i.e., $2^d P$ is constant).

For further discussion of non-classical polynomials, see [**Ta2009**, §1.12]. Henceforth we shall avoid this technical issue by restricting to the high characteristic case $p > d$ (or equivalently, the low degree case $d < p$).

1.4.2. Equidistribution. Let us now consider the equidistribution theory of a classical polynomial $P\colon V \to \mathbf{F}$, where we think of $\mathbf{F}$ as being a fixed field (in particular, $p = O(1)$), and the dimension of V as being very large; V will play the role here that the interval $[N]$ played in Section 1.1. This theory is classical for linear and quadratic polynomials. The general theory was studied first in [**GrTa2009**] in the high characteristic case $p > d$, and extended to the low characteristic case in [**KaLo2008**]; see also [**HaSh2010**], [**HaLo2010**] for some recent refinements. An analogous theory surely exists for the non-classical case, although this is not currently in the literature.

The situation here is simpler because a classical polynomial can only take p values, so that in the equidistributed case one expects each value to be obtained about $|V|/p$ times. Inspired by this, let us call a classical polynomial P *δ-equidistributed* if one has
$$|\{x \in V : P(x) = a\}| - |V|/p| \leq \delta|V|$$
for all $a \in \mathbf{F}$.

---
[11]We thank Andrew Granville for showings us this argument.

1.4. Equidistribution in finite fields

Exercise 1.4.5. Show that this is equivalent to the notion of δ-equidistribution given in Section 1.1, if one gives $\mathbf{F}$ the metric induced from $\mathbf{R}/\mathbf{Z}$, and if one is willing to modify δ by a multiplicative factor depending on p in the equivalences.

Before we study equidistribution in earnest, we first give a classical estimate.

Exercise 1.4.6 (Chevalley-Warning theorem). Let V be a finite dimensional space, and let $P\colon V \to \mathbf{F}$ be a classical polynomial of degree less than $(p-1)\dim(V)$. Show that $\sum_{x \in V} P(x) = 0$. (*Hint:* Identify V with $\mathbf{F}^n$ for some n and apply Exercise 1.4.1. Use the fact that $\sum_{x \in \mathbf{F}} x^i = 0$ for all $1 \leq i < p-1$, which can be deduced by using a change of variables $x \mapsto bx$.) If, furthermore, P has degree less than $\dim(V)$, conclude that for every $a \in \mathbf{F}$, that $|\{x \in V : P(x) = a\}|$ is a multiple of p. (*Hint:* Apply Fermat's little theorem to the quantity $(P-a)^{p-1}$.) In particular, if $x_0 \in V$, then there exists at least one further $x \in V$ such that $P(x) = P(x_0)$.

If P has degree at most d and $x_0 \in V$, obtain the recurrence inequality

$$|\{x \in V : P(x) = P(x_0)\}| \gg_{p,d} |V|.$$

(*Hint:* Normalise $x_0 = 0$, then average the previous claim over all subspaces of V of a certain dimension.)

The above exercise goes some way towards establishing equidistribution, by showing that every element in the image of P is attained a fairly large number of times. But additional techniques will be needed (together with additional hypotheses on P) in order to obtain full equidistribution. It will be convenient to work in the ultralimit setting. Define a *limit classical polynomial* $P\colon V \to \mathbf{F}$ on a limit finite-dimensional vector space $V = \prod_{\alpha \to \alpha_\infty} V_\alpha$ of degree at most d to be an ultralimit of classical polynomials $P_\alpha\colon V_\alpha \to \mathbf{F}$ of degree at most d (we keep $\mathbf{F}$ and d fixed independently of α). We say that a limit classical polynomial P is *equidistributed* if one has

$$|\{x \in V : P(x) = a\}| = |V|/p + o(|V|)$$

for all $a \in \mathbf{F}$, where the cardinalities here are of course limit cardinalities.

Exercise 1.4.7. Let V be a limit finite-dimensional vector space. Show that a limit function $P\colon V \to \mathbf{F}$ is a limit classical polynomial of degree at most d if and only if it is a classical polynomial of degree at most d (observing here that every limit vector space is automatically a vector space).

Exercise 1.4.8. Let $P = \lim_{\alpha \to \alpha_\infty} P_\alpha$ be a limit classical polynomial. Show that P is equidistributed if and only if, for every $\delta > 0$, P_α is δ-equidistributed for α sufficiently close to α_∞.

Exercise 1.4.9. Let $P \colon V \to \mathbf{F}$ be a limit classical polynomial which is linear (i.e., of degree at most 1). Show that P is equidistributed if and only if P is non-constant.

There is an analogue of the Weyl equidistribution criterion in this setting. Call a limit function $P \colon V \to \mathbf{F}$ *biased* if $|\mathbf{E}_{x \in V} e(P(x))| \gg 1$, and *unbiased* if $\mathbf{E}_{x \in V} e(P(x)) = o(1)$, where we identify $P(x) \in \mathbf{F}$ with an element of $\mathbf{R}/\mathbf{Z}$.

Exercise 1.4.10 (Weyl equidistribution criterion). Let $P \colon V \to \mathbf{F}$ be a limit function. Show that P is equidistributed if and only if kP is unbiased for all non-zero $k \in \mathbf{F}$.

Thus to understand the equidistribution of polynomials, it suffices to understand the size of exponential sums $\mathbf{E}_{x \in V} e(P(x))$. For linear polynomials, this is an easy application of Fourier analysis:

Exercise 1.4.11. Let $P \colon V \to \mathbf{F}$ be a polynomial of degree at most 1. Show that $|\mathbf{E}_{x \in V} e(P(x))|$ equals 1 if P is constant, and equals 0 if P is not constant. (Note that this is completely consistent with the previous two exercises.)

Next, we turn our attention to the quadratic case. Here, we can use the Weyl differencing trick, which we phrase as an identity

$$(1.35) \qquad |\mathbf{E}_{x \in V} f(x)|^2 = \mathbf{E}_{h \in V} \mathbf{E}_{x \in V} \Delta_h f(x)$$

for any finite vector space V and function $f \colon V \to \mathbf{C}$, where $\Delta_h f(x) := f(x+h)\overline{f(x)}$ is the multiplicative derivative. Taking ultralimits, we see that the identity also holds for limit functions on limit finite dimensional vector spaces. In particular, we have

$$(1.36) \qquad |\mathbf{E}_{x \in V} e(P(x))|^2 = \mathbf{E}_{h \in V} \mathbf{E}_{x \in V} e(\partial_h P(x))$$

for any limit function $P \colon V \to \mathbf{F}$ on a limit finite dimensional space.

If P is quadratic, then $\partial_h P$ is linear. Applying (1.4.11), we conclude that if P is biased, then $\partial_h P$ must be constant for $\gg |V|$ values of $h \in V$.

On the other hand, by using the cocycle identity

$$\partial_{h+k} P(x) = \partial_h P(x+k) + \partial_k P(x)$$

we see that the set of $h \in V$ for which $\partial_h P$ is constant is a limit subspace of W. On that subspace, P is then linear; passing to a codimension one subspace W' of W, P is then constant on W'. As $\partial_h P$ is linear for every h, P is then linear on each coset $h + W'$ of W'. As $|W'| \gg |V|$, there are only a bounded number of such cosets; thus P is piecewise linear, and thus piecewise constant on slightly smaller cosets. Intersecting all the subspaces together, we can thus find another limit subspace U with $|U| \gg |V|$ such

1.4. Equidistribution in finite fields

that P is constant on each coset of U. To put it another way, if we view U as the intersection of a bounded number of kernels of linear homomorphism $L_1, \ldots, L_d \colon V \to \mathbf{F}$ (where $d = O(1)$ is the codimension of U), then P is constant on every simultaneous level set of $L_1, \ldots, L_d$, and can thus be expressed as a function $F(L_1, \ldots, L_d)$ of these linear polynomials.

More generally, let us say that a limit classical polynomial P of degree $\leq d$ is *low rank* if it can be expressed as $P = F(Q_1, \ldots, Q_d)$ where $Q_1, \ldots, Q_d$ are a bounded number of polynomials of degree $\leq d-1$. We can summarise the above discussion (and also Exercise 1.4.11) as follows:

Proposition 1.4.3. *Let $d \leq 2$, and let $P \colon V \to \mathbf{F}$ be a limit classical polynomial. If P is biased, then P is low rank.*

In particular, from the Weyl criterion, we see that if P is not equidistributed, then P is of low rank.

Of course, the claim fails if the low rank hypothesis is dropped. For instance, consider a limit classical quadratic $Q = L_1 L_2$ that is the product of two linearly independent linear polynomials L_1, L_2. Then Q attains each non-zero value with a density of $(p-1)/p^2$ rather than $1/p$ (and attains 0 with a density of $(2p-1)/p^2$ rather than $1/p$).

Exercise 1.4.12. Suppose that the characteristic p of $\mathbf{F}$ is greater than 2, and suppose that $P \colon \mathbf{F}^n \to \mathbf{F}$ is a quadratic polynomial of the form $P(x) = x^T M x + b^T x + c$, where $c \in \mathbf{F}$, $b \in \mathbf{F}^n$, M is a symmetric $n \times n$ matrix with coefficients in $\mathbf{F}$, and x^T is the transpose of x. Show that $|\mathbf{E}_{x \in V} e(P(x))| \leq p^{-r/2}$, where r is the rank of M. Furthermore, if b is orthogonal to the kernel of M, show that equality is attained, and otherwise $\mathbf{E}_{x \in V} e(P(x))$ vanishes.

What happens in the even characteristic case (assuming now that M is not symmetric)?

Exercise 1.4.13 (Van der Corput lemma). Let $P \colon V \to \mathbf{F}$ be a limit function on a limit finite dimensional vector space V, and suppose that there exists a limit subset H of V which is sparse in the sense that $|H| = o(|V|)$, and such that $\partial_h P$ is equidistributed for all $h \in V \setminus H$. Show that P itself is equidistributed. Use this to give an alternate proof of Proposition 1.4.3.

Exercise 1.4.14 (Space of polynomials is discrete). Let $P \colon V \to \mathbf{F}$ be a polynomial of degree at most d such that $\mathbf{E}_{x \in V} |e(P(x)) - c| < 2^{-d+1}$ for some constant $c \in S^1$. Show that P is constant. (*Hint:* Induct on d.) Conclude that if P, Q are two distinct polynomials of degree at most d, that $\|e(P) - e(Q)\|_{L^2(V)} \gg 1$.

The fact that high rank polynomials are equidistributed extends to higher degrees also:

Theorem 1.4.4. *Let $P\colon V \to \mathbf{F}$ be a limit classical polynomial. If P is biased, then P is low rank.*

In particular, from the Weyl criterion, we see that if P is not equidistributed, then P is of low rank.

In the high characteristic case $p > d$, this claim was shown in [**GrTa2009**]; the generalisation to the low characteristic case $p \leq d$ was carried out in [**KaLo2008**]. The statement is phrased in the language of ultrafilters, but it has an equivalent finitary analogue:

Exercise 1.4.15. Show that Theorem 1.4.4 is equivalent to the claim that for every $d \geq 1$ and $\delta > 0$, and every classical polynomial $P\colon V \to \mathbf{F}$ of degree at most d on a finite-dimensional vector space with $|\mathbf{E}_{x\in V} e(P(x))| \geq \delta$, that P can be expressed as a function of at most $O_{\delta,d}(1)$ classical polynomials of degree at most $d - 1$.

The proof of Theorem 1.4.4 is a little lengthy. It splits up into two pieces. We say that a limit function $P\colon V \to \mathbf{F}$ (not necessarily a polynomial) is *of order $< d$* if it can be expressed as a function of a bounded number of polynomials of degree less than d. Our task is thus to show that every polynomial of degree d which is biased, is of order $< d$. We first get within an epsilon of this goal, using an argument of Bogdanov and Viola [**BoVi2010**]:

Lemma 1.4.5 (Bogdanov-Viola lemma). *Let $P\colon V \to \mathbf{F}$ be a limit polynomial of degree d which is biased, and let $\varepsilon > 0$ be standard. Then one can find a limit function $Q\colon V \to \mathbf{F}$ of order $< d$ such that $|\{x \in V : P(x) \neq Q(x)\}| \leq \varepsilon |V|$.*

Proof. Let $\kappa > 0$ be a small standard number (depending on ε) to be chosen later, let M be a large standard integer (depending on ε, κ) to be chosen later, and let $h_1, \ldots, h_M$ be chosen uniformly at random from V. An application of the second moment method (which we leave as an exercise) shows that if M is large enough, then with probability at least $1 - \varepsilon$, one has
$$|\mathbf{E}_{m\in M} e(P(x + h_m)) - \mathbf{E}_{x\in V} e(P(x))| \leq \kappa$$
for at least $(1 - \varepsilon)|V|$ choices of x. We can rearrange this as
$$|e(P(x)) - \frac{1}{\delta}\mathbf{E}_{m\in M} e(-\partial_{h_m} P(x))| \leq \kappa/\delta$$
where $\delta := |\mathbf{E}_{x\in V} e(P(x))|$; note from hypothesis that $\delta \gg 1$. If we let $F(x)$ be the nearest p^{th} root of unity to $\frac{1}{\delta}\mathbf{E}_{m\in M} e(-\partial_{h_m} P(x))$, then (if κ is small enough) we conclude that $e(P(x)) = F(x)$ for at least $(1 - \varepsilon)|V|$ choices of x. On the other hand, F is clearly of order $< d$, and the claim follows. $\square$

Exercise 1.4.16. Establish the claim left as an exercise in the above proof.

1.4. Equidistribution in finite fields

To conclude the proof of Theorem 1.4.4 from Lemma 1.4.5, it thus suffices to show

Proposition 1.4.6 (Rigidity). *Let $P\colon V \to \mathbf{F}$ be a limit polynomial of degree d which is equal to a limit function $Q\colon V \to \mathbf{F}$ of order $< d$ on at least $1 - \varepsilon$ of V, where $\varepsilon > 0$ is standard. If ε is sufficiently small with respect to d, then P is also of order $< d$.*

This proposition is somewhat tricky to prove, even in the high characteristic case $p > d$. We fix $d < p$ and assume inductively that the proposition (and hence) Theorem 1.4.4 has been demonstrated for all smaller values of d.

The main idea here is to start with the "noisy polynomial" Q, and perform some sort of "error correction" on Q to recover P; the key is then to show that this error correction procedure preserves the property of being order $< d$. From Exercise 1.4.14 we know that *in principle*, this error correction is possible if ε is small enough; but in order to preserve the order $< d$ property we need a more explicit error correction algorithm which is tractable for analysis. This is provided by the following lemma.

Lemma 1.4.7 (Error correction of polynomials). *Let $P\colon V \to \mathbf{F}$ be a (limit) classical polynomial of degree at most d, and let $Q\colon V \to \mathbf{F}$ be a (limit) function which agrees with P at least $1-\varepsilon$ of the time for some $\varepsilon \leq 2^{-d-2}$. Then for every $x \in V$, $P(x)$ is equal to the most common value (i.e., the mode) of $\sum_{\omega \in \{0,1\}^{d+1}\setminus\{0\}} (-1)^{|\omega|-1} Q(x + \omega_1 h_1 + \cdots + \omega_{d+1} h_{d+1})$ as $h_1, \ldots, h_{d+1}$ vary in V.*

Proof. As P is a polynomial of degree at most d, one has
$$\partial_{h_1} \ldots \partial_{h_{d+1}} P(x) = 0$$
for all $x, h_1, \ldots, h_{d+1} \in V$. We rearrange this as
$$P(x) = \sum_{\omega \in \{0,1\}^{d+1}\setminus\{0\}} (-1)^{|\omega|-1} P(x + \omega_1 h_1 + \cdots + \omega_{d+1} h_{d+1}).$$
We conclude that
$$(1.37) \quad P(x) = \sum_{\omega \in \{0,1\}^{d+1}\setminus\{0\}} (-1)^{|\omega|-1} Q(x + \omega_1 h_1 + \cdots + \omega_{d+1} h_{d+1} \omega_{d+1})$$
holds unless P and Q differ at $x + h_1 \omega_1 + \cdots + h_{d+1} \omega_{d+1}$ for some $\omega \in \{0,1\}^{d+1}\setminus\{0\}$.

On the other hand, if x is fixed and $h_1, \ldots, h_{d+1}$ are chosen independently and uniformly at random from V, then for each $\omega \in \{0,1\}^{d+1}\setminus\{0\}$, $x + h_1 \omega_1 + \cdots + h_{d+1} \omega_{d+1}$ is also uniformly distributed in V, and so the probability that P and Q differ at $x + h_1 \omega_1 + \cdots + h_{d+1} \omega_{d+1}$ is at most

2^{-d-2}. Applying the union bound for the $2^{d+1} - 1 < 2^{d+1}$ values of ω under consideration, we conclude that (1.37) happens more than half the time, and the claim follows. $\square$

Note that the above argument in fact shows that the mode is attained for at least $1 - 2^{d+1}\varepsilon$ of the choices of $h_1, \ldots, h_{d+1}$.

In view of this lemma, the goal is now to show that if Q is of order $< d$ and is sufficiently close to a polynomial of degree d, then the mode of $\sum_{\omega \in \{0,1\}^{d+1} \setminus \{0\}} (-1)^{|\omega|-1} Q(x + \omega_1 h_1 + \cdots + \omega_{d+1} h_{d+1})$ is also of order $< d$.

By hypothesis, we have $Q = F(R_1, \ldots, R_m)$ for some standard m and some polynomials $R_1, \ldots, R_m$ of degree $d - 1$. To motivate the general argument, let us first work in an easy model case, in which the $R_1, \ldots, R_m$ are polynomials of degree $d - 1$ that are linearly independent modulo low rank (i.e., order $< d - 2$) errors, i.e., no non-trivial linear combination of $R_1, \ldots, R_m$ over $\mathbf{F}$ is of low rank. This is not the most general case, but is somewhat simpler and will serve to illustrate the main ideas.

The linear independence, combined with the inductive hypothesis, implies that any non-trivial linear combination of $R_1, \ldots, R_m$ is unbiased. From this and Fourier analysis, we see that $\vec{R} := (R_1, \ldots, R_m)$ is jointly equidistributed, thus, in particular, we have

(1.38) $$|S_r| = (p^{-m} + o(1))|V|$$

for all $r \in \mathbf{F}^m$, where $S_r := \{x \in V : \vec{R}(x) = r\}$.

In fact, we have a much stronger equidistribution property than this; not only do we understand the distribution of $\vec{R}(x)$ for a single x, but more generally we can control the distribution of an entire parallelopiped

$$\vec{R}^{[D]}(x, h_1, \ldots, h_D) := (\vec{R}(x + \omega_1 h_1 + \cdots + \omega_D h_D))_{\omega_1, \ldots, \omega_D \in \{0,1\}}$$

for any standard integer $D \geq 0$. Because all the components $\vec{R}$ are polynomials of degree $d - 1$, the quantity $\vec{R}^{[D]}(x, h_1, \ldots, h_D)$ is constrained to the space $\Sigma^{[D]}$, defined as the subspace of $(\mathbf{F}^m)^{2^D}$ consisting of all tuples $r = (r_\omega)_{\omega \in \{0,1\}^D}$ obeying the constraints[12]

$$\sum_{\omega \in F} (-1)^{|\omega|} r_\omega = 0$$

for all faces $F \subset \{0,1\}^D$ of dimension d, where $|\omega| := \omega_1 + \cdots + \omega_D$ is the sign of ω.

---

[12] These constraints are of course vacuous if $D < d$.

1.4. Equidistribution in finite fields

Proposition 1.4.8. $\vec{R}^{[D]}$ is equidistributed in $\Sigma^{[D]}$, thus
$$|\{(x, h_1, \ldots, h_D) \in V^{d+1} : \vec{R}^{[D]}(x, h_1, \ldots, h_D) = r\}|$$
$$= \left(\frac{1}{|\Sigma^{[D]}|} + o(1)\right) |V|^{d+1}$$
for all $r \in \Sigma^{[D]}$. Furthermore, we have the refined bound
$$|\{(h_1, \ldots, h_D) \in V^d : \vec{R}^{[D]}(x, h_1, \ldots, h_D) = r\}|$$
$$= \left(\frac{p^m}{|\Sigma^{[D]}|} + o(1)\right) |V|^d$$
for all $r \in \Sigma^{[D]}$ and all $x \in S_{r_0}$.

Proof. It suffices to prove the second claim. Fix x and $r = (r_\omega)_{\omega \in \{0,1\}^D}$. From the definition of $\Sigma^{[D]}$, we see that r is uniquely determined by the component r_0 and $r_{<d} := (r_\omega)_{\omega \in \{0,1\}^D : 0 < |\omega| < d}$. It will thus suffice to show that
$$|\{(x, h_1, \ldots, h_D) \in V^d : \vec{R}^{[D]}_{<d}(x, h_1, \ldots, h_D) = r_{<d}\}|$$
$$= \left(\frac{p^m}{|\Sigma^{[D]}|} + o(1)\right) |V|^d$$
for all $r_{<d} \in (\mathbf{F}^m)^{\{\omega \in \{0,1\}^D : 0 < |\omega| < d\}}$, where
$$\vec{R}^{[D]}_{<d}(x, h_1, \ldots, h_D)$$
$$:= (\vec{R}(x + \omega_1 h_1 + \cdots + \omega_D h_D))_{\omega \in \{0,1\}^D : 0 < |\omega| < d}.$$
By Fourier analysis, it suffices to show that
$$\mathbf{E}_{h_1, \ldots, h_D \in V} e\left(\xi \cdot \vec{R}^{[D]}_{<d}(x, h_1, \ldots, h_D)\right) = o(1)$$
for any non-zero $\xi \in (\mathbf{F}^m)^{\{\omega \in \{0,1\}^D : 0 < |\omega| < d\}}$. In other words, we need to show that
$$(1.39) \quad \mathbf{E}_{h_1, \ldots, h_D \in V} e\left(\sum_{\omega \in \{0,1\}^D : |\omega| < d} \xi_\omega \cdot \vec{R}(x + \omega_1 h_1 + \cdots + \omega_D h_D)\right) = o(1)$$
whenever the $\xi_\omega \in \mathbf{F}^m$ for $\omega \in \{0,1\}^D, 0 < |\omega| < d$ are not all zero.

Let ω_0 be such that $\xi_{\omega_0} \neq 0$, and such that $|\omega|$ is as large as possible; let us write $d' := |\omega_0|$, so that $0 \le d' < d$. Without loss of generality, we may take $\omega_0 = (1, \ldots, 1, 0, \ldots, 0)$. Suppose (1.39) failed, then by the pigeonhole principle one can find $h_{d'+1}, \ldots, h_D$ such that
$$|\mathbf{E}_{h_1, \ldots, h_{d'} \in V} e(\sum_{\omega \in \{0,1\}^D : |\omega| < d} \xi_\omega \cdot \vec{R}(x + \omega_1 h_1 + \cdots + \omega_D h_D))| \gg 1.$$

We write the left-hand side as

$$\left|\mathbf{E}_{h_1,\ldots,h_{d'}\in V} e(\xi_{\omega_0}\cdot \vec{R}(x+h_1+\cdots+h_{d'}))\prod_{j=1}^{d'} f_j(h_1,\ldots,h_{d'})\right|$$

where f_j are bounded limit functions depending on $x, h_{d'+1},\ldots,h_D$ that are independent of h_j.

We can eliminate each f_j term in turn by the Cauchy-Schwarz argument used in Section 1.3, and conclude that

$$\|e(\xi_{\omega_0}\cdot \vec{R})\|_{U^{d'}(V)} \gg 1,$$

and thus by the monotonicity of Gowers norms

$$\|e(\xi_{\omega_0}\cdot \vec{R})\|_{U^{d-1}(V)} \gg 1$$

or, in other words, that the degree $d-1$ polynomial $(x,h_1,\ldots,h_{d-1})\mapsto \partial_{h_1}\cdots\partial_{h_{d-1}}(\xi_{\omega_0}\cdot \vec{R})(x)$ is biased. By the induction hypothesis, this polynomial must be low rank.

At this point we crucially exploit the high characteristic hypothesis by noting the Taylor expansion formula

$$P(y) = \frac{1}{(d-1)!}\partial_y^{d-1}P(y) + \text{low rank errors}.$$

The high characteristic is necessary here to invert $(d-1)!$. We conclude that $\xi_{\omega_0}\cdot \vec{R}$ is of low rank, but this contradicts the hypothesis on the $R_1,\ldots,R_m$ and the non-zero nature of ξ_{ω_0}, and the claim follows. $\square$

Let $x \in V$ and $r = (r_\omega)_{\omega\in\{0,1\}^D} \in \Sigma^{[D]}$. From the above proposition we have an equidistribution result for a cube pinned at x:

(1.40)
$$|\{(h_1,\ldots,h_D)\in V^D : x+\omega_1 h_1+\cdots+\omega_D h_D \in S_{r_\omega}$$
$$\text{for all } \omega\in\{0,1\}^D\}|$$
$$=\left(\frac{p^m}{|\Sigma^{[D]}|}+o(1)\right)|V|^D.$$

In fact, we can do a bit better than this, and obtain equidistribution even after fixing a second vertex:

Exercise 1.4.17 (Equidistribution of doubly pinned cubes). Let $(r_\omega)_{\omega\in\{0,1\}^D} \in \Sigma^{[D]}$, let $x \in S_{r_0}$, and let $\omega' \in \{0,1\}^D\setminus\{0\}$. Then for all but $o(|V|)$ elements y of $S_{r_{\omega_0}}$, one has

(1.41)
$$|\{(h_1,\ldots,h_D)\in V^D : x+\omega_1 h_1+\cdots+\omega_D h_D \in S_{r_\omega}$$
$$\text{for all } \omega\in\{0,1\}^D; x+\omega'_1 h_1+\cdots+\omega'_D h_D = y\}|$$
$$=(\frac{p^m}{|\Sigma^{[D]}|}+o(1))|V|^{D-1}.$$

1.4. Equidistribution in finite fields

(*Hint:* One can proceed by applying Proposition 1.4.8 with D replaced by a larger dimension, such as $2D$; details can be found in [**GrTa2009**].)

We can now establish Proposition 1.4.6 in the case where the $R_1, \ldots, R_m$ are independent modulo low rank errors. Let $r_0 \in \mathbf{F}^m$ and $x \in S_{r_0}$. It will suffice to show that $P(x)$ does not depend on x as long as x stays inside r_0.

Call an atom S_r *good* if P and Q agree for at least $1 - \sqrt{\varepsilon}$ of the elements of S_r; by Markov's inequality (and (1.38)) we see that at least $1 - \sqrt{\varepsilon} + o(1)$ of the atoms are good. From this and an easy counting argument we can find an element $r = (r_\omega)_{\omega \in \{0,1\}^d}$ in $\Sigma^{[d]}$ with the specified value of r_0, such that r_ω is good for every $\{0,1\}^d \setminus \{0\}$.

Fix r. Now consider all pinned cubes $(x + h_1\omega_1 + \cdots + h_d\omega_d)_{\omega_1, \ldots, \omega_d \in \{0,1\}^d}$ with $x + h_1\omega_1 + \cdots + h_d\omega_d \in S_{r_\omega}$ for all $\omega \in \{0,1\}^d \setminus \{0\}$. By (1.40), the number of such cubes is $(\frac{p^m}{|\Sigma^{[d]}|} + o(1))|V|^d$. On the other hand, by Exercise 1.4.17, the total number of such cubes for which

$$P(x + h_1\omega_1 + \cdots + h_d\omega_d) \neq Q(x + h_1\omega_1 + \cdots + h_d\omega_d)$$

for some $\omega \in \{0,1\}^d \setminus \{0\}$ is $o(|V|^{d-1})$. We conclude that there exists a pinned cube for which

$$P(x + h_1\omega_1 + \cdots + h_d\omega_d) = Q(x + h_1\omega_1 + \cdots + h_d\omega_d)$$

for all $\omega \in \{0,1\}^d \setminus \{0\}$, and, in particular, (1.37) holds. However, as Q is constant on each of the S_r, we see that the right-hand side of (1.37) does not depend on x, and so the same holds true for the left-hand side.

This completes the proof of Proposition 1.4.6 in the independent case. In the general case, one reduces to a (slight generalisation of) this case by the following regularity lemma:

Lemma 1.4.9 (Regularity lemma). *Let $R_1, \ldots, R_m$ be a bounded number of limit classical polynomials of degree $\leq d - 1$. Then there exists a limit classical bounded number of polynomials $S_{d',1}, \ldots, S_{d',m_{d'}}$ of degree $\leq d'$ for each $1 \leq d' \leq d - 1$, such that each $R_1, \ldots, R_m$ is a function of the $S_{d',i}$ for $1 \leq d' \leq d$ and $1 \leq i \leq m_{d'}$, and such that for each d', the $S_{d',1}, \ldots, S_{d',m_{d'}}$ are independent modulo low rank polynomials of degree d'.*

Proof. We induct on d. The claim is vacuously true for $d = 1$, so suppose that $d > 1$ and that the claim has already been proven for $d - 1$.

Let Poly_{d-1} be the space of limit classical polynomials of degree $\leq d - 1$, and let Poly^0_{d-1} be the subspace of low rank limit classical polynomials. Working in the quotient space $\text{Poly}_{d-1}/\text{Poly}^0_{d-1}$, we see that

$R_1, \ldots, R_m$ generates a finite-dimensional space here, which thus has a basis $S_{d-1,1}, \ldots, S_{d-1,m_{d-1}}$ mod Poly_{d-1}^0, thus $S_{d-1,1}, \ldots, S_{d-1,m_{d-1}}$ are linearly independent modulo low rank polynomials of degree $d-1$, and the $R_1, \ldots, R_m$ are linear combinations of the $S_{d-1,1}, \ldots, S_{d-1,m_{d-1}}$ plus combinations of some additional polynomials $R'_1, \ldots, R'_{m'}$ of degree $d-2$. Applying the induction hypothesis to those additional polynomials, one obtains the claim. □

Exercise 1.4.18. Show that the polynomials $S := (S_{d',i})_{1 \leq d' \leq d-1; 1 \leq i \leq m_{d'}}$ appearing in the above lemma are equidistributed in the sense that

$$|\{x \in V : S(x) = s\}| = \left(\frac{1}{p^{\sum_{d'=1}^d m_{d'}}} + o(1) \right) |V|$$

for any $s = (s_{d',i})_{1 \leq d' \leq d-1; 1 \leq i \leq m_{d'}}$ with $s_{d',i} \in \mathbf{F}$.

Applying the above lemma, one can express any order $< d$ function Q in the form $Q = F((S_{d',i})_{1 \leq d' \leq d-1; 1 \leq i \leq m_{d'}})$. It is then possible to modify the previous arguments to obtain Proposition 1.4.6; see [**GrTa2009**] for more details. (We phrase the arguments in a finitary setting rather than a non-standard one, but the two approaches are equivalent; see Section 2.1 for more discussion.)

It is possible to modify the above arguments to handle the low characteristic case, but due to the lack of a good Taylor expansion, one has to regularise the derivatives of the polynomials, as well as the polynomials themselves; see [**KaLo2008**] for details.

1.4.3. Analytic rank. Define the *rank* $\mathrm{rank}_{d-1}(P)$ of a degree d (limit) classical polynomial P to be the least number m of degree $\leq d-1$ (limit) classical polynomials $R_1, \ldots, R_m$ such that P is a function of $R_1, \ldots, R_m$. Proposition 1.4.3 tells us that P is equidistributed whenever the rank is unbounded. However, the proof was rather involved. There is a more elementary approach to equidistribution to Gowers and Wolf [**GoWo2010b**] which replaces the rank by a different object, called *analytic rank*, and which can serve as a simpler substitute for the concept of rank in some applications.

Definition 1.4.10 (Analytic rank). The analytic rank $\mathrm{arank}_{d-1}(P)$ of a (limit) classical polynomial $P \colon V \to \mathbf{F}$ of degree $\leq d$ is defined to be the quantity

$$\mathrm{arank}_d(P) := -\log_p \mathbf{E}_{x,h_1,\ldots,h_d \in V} e(\partial_{h_1} \ldots \partial_{h_d} P(x))$$
$$= -2^d \log_p \|e(P)\|_{U^d(V)}.$$

From the properties of the Gowers norms we see that this quantity is non-negative, is zero if and only if P is a polynomial of degree $< d$, and is

1.4. Equidistribution in finite fields

finite (or limit finite) for $d > 2$. (For $d = 1$, the analytic rank is infinite if P is non-constant and zero if P is constant.)

Exercise 1.4.19. Show that if $p > 2$ and P is a (limit) classical polynomial of degree 2, then $\mathrm{rank}_1(P) = \mathrm{arank}_1(P)$.

Exercise 1.4.20. Show that if the analytic rank $\mathrm{arank}_{d-1}(P)$ of a limit classical polynomial P of degree d is unbounded, then P is equidistributed.

Exercise 1.4.21. Suppose we are in the high characteristic case $p > d$. Using Proposition 1.4.3, show that a limit classical polynomial has bounded analytic rank if and only if it has bounded rank. (*Hint*: One direction follows from the preceding exercise. For the other direction, use the Taylor formula $P(x) = \frac{1}{d!}\partial_x^d P(x)$.) This is a special case of the *inverse conjecture for the Gowers norms*, which we will discuss in more detail in later sections.

Conclude the following finitary version: if $P\colon V \to \mathbf{F}$ is a classical polynomial of degree d on a finite-dimensoinal vector space V, and $\mathrm{arank}_{d-1}(P) \leq M$, then $\mathrm{rank}_{d-1}(P) \ll_{M,p,d} 1$; conversely, if $\mathrm{rank}_{d-1}(P) \leq M$, then $\mathrm{arank}_{d-1}(P) \ll_{M,p,d} 1$.

Exercise 1.4.22. Show that if P is a (limit) classical polynomial of degree d, then $\mathrm{rank}_{d-1}(P) = \mathrm{rank}_{d-1}(cP)$ and $\mathrm{arank}_{d-1}(P) = \mathrm{arank}_{d-1}(cP)$ for all $c \in \mathbf{F}\backslash 0$, and $\mathrm{rank}_{d-1}(P+Q) = \mathrm{rank}_{d-1}(P)$ and $\mathrm{arank}_{d-1}(P+Q) = \mathrm{arank}_{d-1}(P)$ for all (limit) classical polynomials Q of degree $\leq d-1$.

It is clear that the rank obeys the triangle inequality $\mathrm{rank}_{d-1}(P+Q) \leq \mathrm{rank}_{d-1}(P) + \mathrm{rank}_{d-1}(Q)$ for all (limit) classical polynomials of degree $\leq d$. There is an analogue for analytic rank:

Proposition 1.4.11 (Quasi-triangle inequality for analytic rank). (*See* [GoWo2010b].) *Let* $P, Q\colon V \to \mathbf{F}$ *be (limit) classical polynomials of degree* $\leq d$. *Then* $\mathrm{arank}_{d-1}(P+Q) \leq 2^d(\mathrm{arank}_{d-1}(P) + \mathrm{arank}_{d-1}(Q))$.

Proof. Let $T_1(h_1, \ldots, h_d)$ be the d-linear form

$$T_1(h_1, \ldots, h_d) := \partial_{h_1} \ldots \partial_{h_d} P(x)$$

(note that the right-hand side is independent of x); similarly define

$$T_2(h_1, \ldots, h_d) := \partial_{h_1} \ldots \partial_{h_d} P(x).$$

By definition, we have

$$\mathbf{E}_{h_1,\ldots,h_d \in V} e(T_1(h_1, \ldots, h_d)) = p^{-\mathrm{arank}_{d-1}(P)}$$

and

$$\mathbf{E}_{h_1,\ldots,h_d \in V} e(T_2(h_1, \ldots, h_d)) = p^{-\mathrm{arank}_{d-1}(Q)}$$

and thus

$$\mathbb{E}_{h_1,\ldots,h_d,h'_1,\ldots,h'_d \in V} e(T_1(h_1,\ldots,h_d) + T_2(h'_1,\ldots,h'_d))$$
$$= p^{-\operatorname{arank}_{d-1}(P) - \operatorname{arank}_{d-1}(Q)}.$$

We make the substitution $h'_j = h_j + k_j$. Using the multilinearity of T_2, we can write the left-hand side as

$$\mathbb{E}_{k_1,\ldots,k_d \in V} \mathbb{E}_{h_1,\ldots,h_d \in V} e((T_1 + T_2)(h_1,\ldots,h_d))$$
$$\times \prod_{j=1}^d f_j(h_1,\ldots,h_d,k_1,\ldots,k_d)$$

where the f_j are functions bounded in magnitude by 1 that are independent of the h_j variable. Eliminating all these factors by Cauchy-Schwarz as in Section 1.3, we can bound the above expression by

$$|\mathbb{E}_{h_1^0,\ldots,h_d^0,h_1^1,\ldots,h_d^1 \in V} e(\sum_{\omega \in \{0,1\}^d} (-1)^{|\omega|}(T_1+T_2)(h_1^{\omega_1},\ldots,h_d^{\omega_d}))|^{1/2^d}$$

which, using the substitution $h_i := h_i^1 - h_i^0$ and the multilinearity of $T_1 + T_2$, simplifies to

$$|\mathbb{E}_{h_1,\ldots,h_d \in V} e((T_1+T_2)(h_1,\ldots,h_d))|^{1/2^d}$$

which by definition of analytic rank is

$$p^{-\operatorname{arank}_{d-1}(P+Q)/2^d},$$

and the claim follows. $\square$

1.5. The inverse conjecture for the Gowers norm I. The finite field case

In Section 1.3, we saw that the number of additive patterns in a given set was (in principle, at least) controlled by *the Gowers uniformity norms* of functions associated to that set.

Such norms can be defined on any finite additive group (and also on some other types of domains, though we will not discuss this point here). In particular, they can be defined on the finite-dimensional vector spaces V over a finite field $\mathbf{F}$.

In this case, the Gowers norms $U^{d+1}(V)$ are closely tied to the space $\operatorname{Poly}_{\leq d}(V \to \mathbf{R}/\mathbf{Z})$ of polynomials of degree at most d. Indeed, as noted in Exercise 1.4.20, a function $f \colon V \to \mathbf{C}$ of $L^\infty(V)$ norm 1 has $U^{d+1}(V)$ norm equal to 1 if and only if $f = e(\phi)$ for some $\phi \in \operatorname{Poly}_{\leq d}(V \to \mathbf{R}/\mathbf{Z})$; thus polynomials solve the "100% inverse problem" for the trivial inequality $\|f\|_{U^{d+1}(V)} \leq \|f\|_{L^\infty(V)}$. They are also a crucial component of the solution

1.5. Inverse conjecture over finite fields

to the "99% inverse problem" and "1% inverse problem". For the former, we will soon show:

Proposition 1.5.1 (99% inverse theorem for $U^{d+1}(V)$). *Let $f\colon V \to \mathbf{C}$ be such that $\|f\|_{L^\infty(V)}$ and $\|f\|_{U^{d+1}(V)} \geq 1 - \varepsilon$ for some $\varepsilon > 0$. Then there exists $\phi \in \mathrm{Poly}_{\leq d}(V \to \mathbf{R}/\mathbf{Z})$ such that $\|f - e(\phi)\|_{L^1(V)} = O_{d,\mathbf{F}}(\varepsilon^c)$, where $c = c_d > 0$ is a constant depending only on d.*

Thus, for the Gowers norm to be almost completely saturated, one must be very close to a polynomial. The converse assertion is easily established:

Exercise 1.5.1 (Converse to 99% inverse theorem for $U^{d+1}(V)$). *If $\|f\|_{L^\infty(V)} \leq 1$ and $\|f - e(\phi)\|_{L^1(V)} \leq \varepsilon$ for some $\phi \in \mathrm{Poly}_{\leq d}(V \to \mathbf{R}/\mathbf{Z})$, then $\|F\|_{U^{d+1}(V)} \geq 1 - O_{d,\mathbf{F}}(\varepsilon^c)$, where $c = c_d > 0$ is a constant depending only on d.*

In the 1% world, one no longer expects to be close to a polynomial. Instead, one expects to *correlate* with a polynomial. Indeed, one has

Lemma 1.5.2 (Converse to the 1% inverse theorem for $U^{d+1}(V)$). *If $f\colon V \to \mathbf{C}$ and $\phi \in \mathrm{Poly}_{\leq d}(V \to \mathbf{R}/\mathbf{Z})$ are such that $|\langle f, e(\phi)\rangle_{L^2(V)}| \geq \varepsilon$, where $\langle f, g\rangle_{L^2(V)} := \mathbf{E}_{x \in G} f(x)\overline{g(x)}$, then $\|f\|_{U^{d+1}(V)} \geq \varepsilon$.*

Proof. From the definition (1.34) of the U^1 norm, the monotonicity of the Gowers norms (Exercise 1.3.19), and the polynomial phase modulation invariance of the Gowers norms (Exercise 1.3.21), one has

$$|\langle f, e(\phi)\rangle| = \|fe(-\phi)\|_{U^1(V)}$$
$$\leq \|fe(-\phi)\|_{U^{d+1}(V)}$$
$$= \|f\|_{U^{d+1}(V)}$$

and the claim follows. □

It is a difficult but known fact that Lemma 1.5.2 can be reversed:

Theorem 1.5.3 (1% inverse theorem for $U^{d+1}(V)$). *Suppose that $\mathrm{char}(\mathbf{F}) > d \geq 0$. If $f\colon V \to \mathbf{C}$ is such that $\|f\|_{L^\infty(V)} \leq 1$ and $\|f\|_{U^{d+1}(V)} \geq \varepsilon$, then there exists $\phi \in \mathrm{Poly}_{\leq d}(V \to \mathbf{R}/\mathbf{Z})$ such that $|\langle f, e(\phi)\rangle_{L^2(V)}| \gg_{\varepsilon,d,\mathbf{F}} 1$.*

This result is sometimes referred to as the *inverse conjecture for the Gowers norm* (in high, but bounded, characteristic). For small d, the claim is easy:

Exercise 1.5.2. Verify the cases $d = 0, 1$ of this theorem. (*Hint:* To verify the $d = 1$ case, use the Fourier-analytic identities $\|f\|_{U^2(V)} = (\sum_{\xi \in \hat{V}} |\hat{f}(\xi)|^4)^{1/4}$

and $\|f\|_{L^2(V)} = (\sum_{\xi \in \hat{V}} |\hat{f}(\xi)|^2)^{1/2}$, where $\hat{V}$ is the space of all homomorphisms $\xi \colon x \mapsto \xi \cdot x$ from V to $\mathbf{R}/\mathbf{Z}$, and $\hat{f}(\xi) := \mathbf{E}_{x \in V} f(x) e(-\xi \cdot x)$ are the Fourier coefficients of f.)

This conjecture for larger values of d are more difficult to establish. The $d = 2$ case of the theorem was established in [**GrTa2008**]; the low characteristic case char$(\mathbf{F}) = d = 2$ was independently and simultaneously established in [**Sa2007**]. The cases $d > 2$ in the high characteristic case was established in two stages: first, using a modification of the Furstenberg correspondence principle in [**TaZi2010**], and then using a modification of the methods of Host and Kra [**HoKr2005**] and Ziegler [**Zi2007**] to solve that counterpart, as done in [**BeTaZi2010**]; an alternate proof was also obtained in [**Sz2010c**]. Finally, the low characteristic case was recently achieved in [**TaZi2011**].

In the high characteristic case, we saw from Section 1.4 that one could replace the space of non-classical polynomials $\mathrm{Poly}_{\leq d}(V \to \mathbf{R}/\mathbf{Z})$ in the above conjecture with the essentially equivalent space of classical polynomials $\mathrm{Poly}_{\leq d}(V \to \mathbf{F})$. However, as we shall see below, this turns out not to be the case in certain low characteristic cases (a fact first observed in [**LoMeSa2008**], [**GrTa2009**]), for instance, if char$(\mathbf{F}) = 2$ and $d \geq 3$; this is ultimately due to the existence in those cases of non-classical polynomials which exhibit no significant correlation with classical polynomials of equal or lesser degree. This distinction between classical and non-classical polynomials appears to be a rather non-trivial obstruction to understanding the low characteristic setting; it may be necessary to obtain a more complete theory of non-classical polynomials in order to fully settle this issue.

The inverse conjecture has a number of consequences. For instance, it can be used to establish the analogue of Szemerédi's theorem in this setting:

Theorem 1.5.4 (Szemerédi's theorem for finite fields). *Let $\mathbf{F} = \mathbf{F}_p$ be a finite field, let $\delta > 0$, and let $A \subset \mathbf{F}^n$ be such that $|A| \geq \delta |\mathbf{F}^n|$. If n is sufficiently large depending on p, δ, then A contains an (affine) line $\{x, x+r, \ldots, x+(p-1)r\}$ for some $x, r \in \mathbf{F}^n$ with $r \neq 0$.*

Exercise 1.5.3. Use Theorem 1.5.4 to establish the following generalisation: with the notation as above, if $k \geq 1$ and n is sufficiently large depending on p, δ, then A contains an affine k-dimensional subspace.

We will prove this theorem in two different ways, one using a density increment method, and the other using an energy increment method. We discuss some other applications below the fold.

1.5.1. The 99% inverse theorem. We now prove Proposition 1.5.1. Results of this type for general d appear in [**AlKaKrLiRo2003**] (see also

1.5. Inverse conjecture over finite fields

[**SuTrVa1999**] for a precursor result); the $d = 1$ case was treated previously in [**BlLuRu1993**]. The argument here is taken from [**TaZi2010**], and has a certain "cohomological" flavour (comparing cocycles with coboundaries, determining when a closed form is exact, etc.). Indeed, the inverse theory can be viewed as a sort of "additive combinatorics cohomology".

Let $\mathbf{F}, V, d, f, \varepsilon$ be as in the theorem. We let all implied constants depend on $d, \mathbf{F}$. We use the symbol c to denote various positive constants depending only on d. We may assume ε is sufficiently small depending on $d, \mathbf{F}$, as the claim is trivial otherwise.

The case $d = 0$ is easy, so we assume inductively that $d \geq 1$ and that the claim has been already proven for $d - 1$.

The first thing to do is to make f unit magnitude. One easily verifies the crude bound
$$\|f\|_{U^{d+1}(V)}^{2^{d+1}} \leq \|f\|_{L^1(V)}$$
and thus
$$\|f\|_{L^1(V)} \geq 1 - O(\varepsilon).$$
Since $|f| \leq 1$ pointwise, we conclude that
$$\mathbf{E}_{x \in V} 1 - |f(x)| = O(\varepsilon).$$
As such, f differs from a function $\tilde{f}$ of unit magnitude by $O(\varepsilon)$ in L^1 norm. By replacing f with $\tilde{f}$ and using the triangle inequality for the Gowers norm (changing ε and worsening the constant c in Proposition 1.5.1 if necessary), we may assume, without loss of generality, that $|f| = 1$ throughout, thus $f = e(\psi)$ for some $\psi \colon V \to \mathbf{R}/\mathbf{Z}$.

Since
$$\|f\|_{U^{d+1}(V)}^{2^{d+1}} = \mathbf{E}_{h \in V} \|e(\partial_h \psi)\|_{U^d(V)}^{2^d}$$
we see from Markov's inequality that
$$\|e(\partial_h \psi)\|_{U^d(V)} \geq 1 - O(\varepsilon^c)$$
for all h in a subset H of V of density $1 - O(\varepsilon^c)$. Applying the inductive hypothesis, we see that for each such h, we can find a polynomial $\phi_h \in \mathrm{Poly}_{\leq d-1}(V \to \mathbf{R}/\mathbf{Z})$ such that
$$\|e(\partial_h \psi) - e(\phi_h)\|_{L^1(V)} = O(\varepsilon^c).$$
Now let $h, k \in H$. Using the cocycle identity
$$e(\partial_{h+k} \psi) = e(\partial_h \phi) T^h e(\partial_k \phi)$$
where T^h is the shift operator $T^h f(x) := f(x + h)$, we see using Hölder's inequality that
$$\|e(\partial_{h+k} \psi) - e(\phi_h T^h \phi_k)\|_{L^1(V)} = O(\varepsilon^c).$$

On the other hand, $\phi_h T^h \phi_k$ is a polynomial of order d. Also, since H is so dense, every element l of V has at least one representation of the form $l = h+k$ for some $h, k \in H$ (indeed, out of all $|V|$ possible representations $l = h+k$, h or k can fall outside of H for at most $O(\varepsilon^c|V|)$ of these representations). We conclude that for every $l \in V$ there exists a polynomial $\phi'_l \in \text{Poly}_{\leq d}(V \to \mathbf{R}/\mathbf{Z})$ such that

$$\|e(\partial_l \psi) - e(\phi'_l)\|_{L^1(V)} = O(\varepsilon^c). \tag{1.42}$$

The new polynomial ϕ'_l supercedes the old one ϕ_l; to reflect this, we abuse notation and write ϕ_l for ϕ'_l. Applying the cocycle equation again, we see that

$$\|e(\phi_{h+k}) - e(\phi_h T^h \phi_k)\|_{L^1(V)} = O(\varepsilon^c) \tag{1.43}$$

for all $h, k \in V$. Applying the rigidity of polynomials (Exercise 1.4.6), we conclude that

$$\phi_{h+k} = \phi_h T^h \phi_k + c_{h,k}$$

for some constant $c_{h,k} \in \mathbf{R}/\mathbf{Z}$. From (1.43) we in fact have $c_{h,k} = O(\varepsilon^c)$ for all $h, k \in V$.

The expression $c_{h,k}$ is known as a *2-coboundary* (see [**Ta2009**, §1.13] for more discussion). To eliminate it, we use the finite characteristic to discretise the problem as follows. First, we use the cocycle identity

$$\prod_{j=0}^{p-1} e(T^{jh} \partial_h \psi) = 1$$

where p is the characteristic of the field. Using (1.42), we conclude that

$$\|\prod_{j=0}^{p-1} e(T^{jh} \phi_h) - 1\|_{L^1(V)} = O(\varepsilon^c).$$

On the other hand, $T^{jh} \phi_h$ takes values in some coset of a finite subgroup C of $\mathbf{R}/\mathbf{Z}$ (depending only on p, d), by Lemma 1.4.1. We conclude that this coset must be a shift of C by $O(\varepsilon^c)$. Since ϕ_h itself takes values in some coset of a finite subgroup, we conclude that there is a finite subgroup C' (depending only on p, d) such that each ϕ_h takes values in a shift of C' by $O(\varepsilon^c)$.

Next, we note that we have the freedom to shift each ϕ_h by $O(\varepsilon^c)$ (adjusting $c_{h,k}$ accordingly) without significantly affecting any of the properties already established. Doing so, we can thus ensure that all the ϕ_h take values in C' itself, which forces $c_{h,k}$ to do so also. But since $c_{h,k} = O(\varepsilon^c)$, we conclude that $c_{h,k} = 0$ for all h, k, thus ϕ_h is a perfect cocycle:

$$\phi_{h+k} = \phi_h T^h \phi_k.$$

1.5. Inverse conjecture over finite fields

We may thus integrate ϕ_h and write $\phi_h = \partial_h \Phi$, where $\Phi(x) := \phi_x(0)$. Thus $\partial_h \Phi$ is a polynomial of degree $d-1$ for each h, thus Φ itself is a polynomial of degree d. From (1.42) one has

$$\mathbf{E}_{x \in V} e(\partial_h(\psi - \Phi)) = 1 + O(\varepsilon^c)$$

for all $h \in V$; averaging in V we conclude that

$$|\mathbf{E}_{x \in V} e(\psi - \Phi)|^2 = 1 + O(\varepsilon^c)$$

and thus

$$\|e(\psi) - e(\Phi)\|_{L^1(V)} = O(\varepsilon^c)$$

and Proposition 1.5.1 follows.

One consequence of Proposition 1.5.1 is that the property of being a classical polynomial of a fixed degree d is *locally testable*, which is a notion of interest in theoretical computer science. More precisely, suppose one is given a large finite vector space V and two functions $\phi_1, \phi_2 \colon V \to \mathbf{F}$. One is told that one of the functions ϕ_1, ϕ_2 is a classical polynomial of degree at most d, while the other is quite far from being such a classical polynomial, in the sense that every polynomial of degree at most d will differ with that polynomial on at least ε of the values in V. The task is then to decide with a high degree of confidence which of the functions is a polynomial and which one is not, without inspecting too many of the values of ϕ_1 or ϕ_2.

This can be done as follows. Pick $x, h_1, \ldots, h_{d+1} \in V$ at random, and test whether the identities

$$\partial_{h_1} \ldots \partial_{h_{d+1}} \phi_1(x) = 0$$

and

$$\partial_{h_1} \ldots \partial_{h_{d+1}} \phi_2(x) = 0$$

hold; note that one only has to inspect ϕ_1, ϕ_2 at 2^{d+1} values in V for this. If one of these identities fails, then that function must not be polynomial, and so one has successfully decided which of the functions is polynomials. We claim that the probability that the identity fails for the non-polynomial function is at least δ for some $\delta \gg_{d,\mathbf{F}} \varepsilon^{O_{d,\mathbf{F}}(1)}$, and so if one iterates this test $O_\delta(1)$ times, one will be able to successfully solve the problem with probability arbitrarily close to 1.

To verify the claim, suppose for contradiction that the identity only failed at most δ of the time for the non-polynomial (say it is ϕ_2); then $\|e(\phi_2)\|_{U^{d+1}(V)} \geq 1 - O(\delta)$, and thus by Proposition 1.5.1, ϕ_2 is very close in L^1 norm to a polynomial; rounding that polynomial to a root of unity we thus see that ϕ_2 agrees with high accuracy to a classical polynomial, which leads to a contradiction if δ is chosen suitably.

1.5.2. A partial counterexample in low characteristic. We now show a distinction between classical polynomials and non-classical polynomials that causes the inverse conjecture to fail in low characteristic if one insists on using classical polynomials. For simplicity we restrict attention to the characteristic two case $\mathbf{F} = \mathbf{F}_2$. We will use an argument of Alon and Beigel [**AlBe2001**], reproduced in [**GrTa2009**]. A different argument (with stronger bounds) appeared independently in [**LoMeSa2008**].

We work in a standard vector space $V = \mathbf{F}^n$, with standard basis $e_1, \ldots, e_n$ and coordinates $x_1, \ldots, x_n$. Among all the classical polynomials on this space are the *symmetric polynomials*

$$S_m := \sum_{1 \le i_1 < \cdots < i_m \le n} x_{i_1} \ldots x_{i_m},$$

which play a special role.

Exercise 1.5.4. Let $L \colon V \to \mathbf{N}$ be the digit summation function $L := \#\{1 \le i \le n : x_i = 1\}$. Show that

$$S_m = \binom{L}{m} \mod 2.$$

Establish *Lucas' theorem*[13]

$$S_m = S_{2^{j_1}} \ldots S_{2^{j_r}}$$

where $m = 2^{j_1} + \cdots + 2^{j_r}$, $j_1 > \cdots > j_r$ is the binary expansion of m. Show that S_{2^j} is the 2^j binary coefficient of L, and conclude that S_m is a function of L mod 2^{j_1}.

We define an *an affine coordinate subspace* to be a translate of a subspace of V generated by some subset of the standard basis vectors $e_1, \ldots, e_n$. To put it another way, an affine coordinate subspace is created by freezing some of the coordinates, but letting some other coordinates be arbitrary.

Of course, not all classical polynomials come from symmetric polynomials. However, thanks to an application of Ramsey's theorem observed in [**AlBe2001**], this is true on coordinate subspaces:

Lemma 1.5.5 (Ramsey's theorem for polynomials). *Let $P \colon \mathbf{F}^n \to \mathbf{F}$ be a polynomial of degree at most d. Then one can partition $\mathbf{F}^n$ into affine coordinate subspaces of dimension W at least $\omega_d(n)$, where $\omega_d(n) \to \infty$ as $n \to \infty$ for fixed d, such that on each such subspace W, P is equal to a linear combination of the symmetric polynomials $S_0, S_1, \ldots, S_d$.*

---

[13]These results are closely related to the well-known fact that *Pascal's triangle* modulo 2 takes the form of an infinite *Sierpinski gasket*.

1.5. Inverse conjecture over finite fields

Proof. We induct on d. The claim is trivial for $d = 0$, so suppose that $d \geq 1$ and the claim has already been proven for smaller d. The degree d term P_d of P can be written as

$$P_d = \sum_{\{i_1,\ldots,i_d\} \in E} x_{i_1} \ldots x_{i_d}$$

where E is a d-uniform *hypergraph* on $\{1, \ldots, n\}$, i.e., a collection of d-element subsets of $\{1, \ldots, n\}$. Applying *Ramsey's theorem* for hypergraphs (see e.g. [GrRoSp1980] or [Ta2009, §2.6]), one can find a subcollection $j_1, \ldots, j_m$ of indices with $m \geq \omega_d(n)$ such that E either has no edges in $\{j_1, \ldots, j_m\}$, or else contains all the edges in $\{j_1, \ldots, j_m\}$. We then foliate $\mathbf{F}^n$ into the affine subspaces formed by translating the coordinate subspace generated by $e_{j_1}, \ldots, e_{j_m}$. By construction, we see that on each such subspace, P is equal to either 0 or S_d plus a polynomial of degree $d - 1$. The claim then follows by applying the induction hypothesis (and noting that the linear span of $S_0, \ldots, S_{d-1}$ on an affine coordinate subspace is equivariant with respect to translation of that subspace). $\square$

Because of this, if one wants to concoct a function which is almost orthogonal to all polynomials of degree at most d, it will suffice to build a function which is almost orthogonal to the symmetric polynomials $S_0, \ldots, S_d$ on all affine coordinate subspaces of moderately large size. Pursuing this idea, we are led to

Proposition 1.5.6 (Counterexample to classical inverse conjecture)**.** *Let $d \geq 1$, and let $f: \mathbf{F}_2^n \to S^1$ be the function $f := e(L/2^d)$, where L is as in Exercise 1.5.4. Then $L/2^d$ mod 1 is a non-classical polynomial of degree at most d, and so $\|f\|_{U^{d+1}(\mathbf{F}_2^n)} = 1$; but one has*

$$\langle f, e(\phi) \rangle_{L^2(\mathbf{F}_2^n)} = o_{n \to \infty; d}(1)$$

uniformly for all classical polynomials ϕ of degree less than 2^{d-1}, where $o_{n \to \infty; d}(1)$ is bounded in magnitude by a quantity that goes to zero as $n \to \infty$ for each fixed d.

Proof. We first prove the polynomiality of $L/2^d$ mod 1. Let $x \mapsto |x|$ be the obvious map from $\mathbf{F}_2$ to $\{0, 1\}$, thus

$$L = \sum_{i=1}^n |x_i|.$$

By linearity, it will suffice to show that each function $|x_i|$ mod 2^d is a polynomial of degree at most d. But one easily verifies that for any $h \in \mathbf{F}_2^n$, $\partial_h |x_i|$ is equal to zero when $h_i = 0$ and equal to $1 - 2|x_i|$ when $h_i = 1$. Iterating this observation d times, we obtain the claim.

Now let ϕ be a classical polynomial of degree less than 2^{d-1}. By Lemma 1.5.5, we can partition $\mathbf{F}_2^n$ into affine coordinate subspaces W of dimension at least $\omega_d(n)$ such that ϕ is a linear combination of $S_0, \ldots, S_{2^{d-1}-1}$ on each such subspace. By the pigeonhole principle, we thus can find such a W such that
$$|\langle f, e(\phi)\rangle_{L^2(\mathbf{F}_2^n)}| \leq |\langle f, e(\phi)\rangle_{L^2(W)}|.$$
On the other hand, from Exercise 1.5.4, the function ϕ on W depends only on L mod 2^{d-1}. Now, as $\dim(W) \to \infty$, the function L mod 2^d (which is essentially the distribution function of a simple random walk of length $\dim(V)$ on $\mathbf{Z}/2^d\mathbf{Z}$) becomes equidistributed; in particular, for any $a \in \mathbf{Z}/2^d\mathbf{Z}$, the function f will take the values $e(a/2^d)$ and $-e(a/2^d)$ with asymptotically equal frequency on W, whilst ϕ remains unchanged. As such we see that $|\langle f, e(\phi)\rangle_{L^2(W)}| \to 0$ as $\dim(W) \to \infty$, and thus as $n \to \infty$, and the claim follows. $\square$

Exercise 1.5.5. With the same setup as the previous proposition, show that $\|e(S_{2^{d-1}}/2)\|_{U^{d+1}(\mathbf{F}_2^n)} \gg 1$, but that $\langle e(S^{2^{d-1}}/2), e(\phi)\rangle_{L^2(\mathbf{F}_2^n)} = o_{n\to\infty;d}(1)$ for all classical polynomials ϕ of degree less than 2^{d-1}.

1.5.3. The 1% inverse theorem: sketches of a proof. The proof of Theorem 1.5.3 is rather difficult once $d \geq 2$; even the $d = 2$ case is not particularly easy. However, the arguments still have the same cohomological flavour encountered in the 99% theory. We will not give full proofs of this theorem here, but indicate some of the main ideas.

We begin by discussing (quite non-rigorously) the significantly simpler (but still non-trivial) $d = 2$ case, under the assumption of odd characteristic, in which case we can use the arguments from [**Go1998**], [**GrTa2008**]. Unsurprisingly, we will take advantage of the $d = 1$ case of the theorem as an induction hypothesis.

Let $V = \mathbf{F}^n$ for some field $\mathbf{F}$ of characteristic greater than 2, and let f be a function with $\|f\|_{L^\infty(V)} \leq 1$ and $\|f\|_{U^3(V)} \gg 1$. We would like to show that f correlates with a quadratic phase function $e(\phi)$ (due to the characteristic hypothesis, we may take ϕ to be classical), in the sense that $|\langle f, e(\phi)\rangle_{L^2(V)}| \gg 1$.

We expand $\|f\|_{U^3(V)}^8$ as $\mathbf{E}_{h \in V}\|\Delta_h f\|_{U^2(V)}^4$. By the pigeonhole principle, we conclude that
$$\|\Delta_h f\|_{U^2(V)} \gg 1$$
for "many" $h \in V$, where by "many" we mean "a proportion of $\gg 1$". Applying the U^2 inverse theorem, we conclude that for many h, that there exists a linear polynomial $\phi_h \colon V \to \mathbf{F}$ (which we may as well take to be classical) such that
$$|\langle \Delta_h f, e(\phi_h)\rangle_{L^2(V)}| \gg 1.$$

1.5. Inverse conjecture over finite fields

This should be compared with the 99% theory. There, we were able to force $\Delta_h f$ close to $e(\phi_h)$ for most h; here, we only have the weaker statement that $\Delta_h f$ *correlates* with $e(\phi_h)$ for *many* (not *most*) h. Still, we will keep going. In the 99% theory, we were able to assume f had magnitude 1, which made the cocycle equation $\Delta_{h+k} f = (\Delta_h f) T^h \Delta_k f$ available; this then forced an approximate cocycle equation $\phi_{h+k} \approx \phi_h + T^h \phi_k$ for most h, k (indeed, we were able to use this trick to upgrade "most" to "all").

This doesn't quite work in the 1% case. First, f need not have magnitude exactly equal to 1. This is not a terribly serious problem, but the more important difficulty is that correlation, unlike the property of being close, is not transitive or multiplicative: just because $\Delta_h f$ correlates with $e(\phi_h)$, and $T^h \Delta_k f$ correlates with $T^h e(\phi_k)$, one cannot then conclude that $\Delta_{h+k} f = (\Delta_h f) T^h \Delta_k f$ correlates with $e(\phi_h) T^h e(\phi_k)$; and even if one had this, and if $\Delta_{h+k} f$ correlated with $e(\phi_{h+k})$, one could not conclude that $e(\phi_{h+k})$ correlated with $e(\phi_h) T^h e(\phi_k)$.

Despite all these obstacles, it is still possible to extract something resembling a cocycle equation for the ϕ_h, by means of the Cauchy-Schwarz inequality. Indeed, we have the following remarkable observation of Gowers [Go1998]:

Lemma 1.5.7. *Let V be a finite additive group, and let $f: V \to \mathbf{C}$ be a function, bounded by 1. Let $H \subset V$ be a subset with $|H| \gg |V|$, and suppose that for each $h \in H$, suppose that we have a function $\chi_h: V \to \mathbf{C}$ bounded by 1, such that*

$$|\langle \Delta_h f, \chi_h \rangle_{L^2(V)}| \gg 1$$

uniformly in h. Then there exist $\gg |V|^3$ quadruples $h_1, h_2, h_3, h_4 \in H$ with $h_1 + h_2 = h_3 + h_4$ such that

$$|\mathbf{E}_{x \in V} \chi_{h_1}(x) \chi_{h_2}(x + h_1 - h_4) \overline{\chi_{h_3}}(x) \overline{\chi_{h_4}}(x + h_1 - h_4)| \gg 1$$

uniformly among the quadruples.

We shall refer to quadruples (h_1, h_2, h_3, h_4) obeying the relation $h_1 + h_2 = h_3 + h_4$ as *additive quadruples*.

Proof. We extend χ_h to be zero when h lies outside of H. Then we have

$$|\mathbf{E}_{h \in V} \theta_h \langle \Delta_h f, \chi_h \rangle_{L^2(V)}| \gg 1$$

and some complex numbers θ_h bounded in magnitude by one. We rearrange this as

$$|\mathbf{E}_{x,y \in V} f(y) \overline{f(x)} \theta_{y-x} \chi_{y-x}(x)| \gg 1.$$

Using Cauchy-Schwarz in x and y to eliminate the f variables, we conclude that

$$|\mathbf{E}_{x,y,x',y' \in V} \theta_{y-x} \overline{\theta_{y-x'}} \overline{\theta_{y'-x}} \theta_{y'-x} \chi_{y-x}(x)$$
$$\overline{\chi_{y-x'}(x')} \chi_{y'-x}(x) \overline{\chi_{y'-x'}(x')}| \gg 1.$$

Setting (h_1, h_2, h_3, h_4) to be the additive quadruple $(y-x, y'-x', y'-x, y-x')$ we obtain

$$|\mathbf{E}_{h_1+h_2=h_3+h_4} \theta_{h_1} \theta_{h_2} \overline{\theta_{h_3} \theta_{h_4}}$$
$$\mathbf{E}_{x \in V} \chi_{h_1}(x) \chi_{h_2}(x+h_1-h_4) \overline{\chi_{h_3}}(x) \overline{\chi_{h_4}}(x+h_1-h_4)| \gg 1$$

and the claim follows (note that for the quadruples obeying the stated lower bound, h_1, h_2, h_3, h_4 must lie in H). $\square$

Applying this lemma to our current situation, we find many additive quadruples (h_1, h_2, h_3, h_4) for which

$$|\mathbf{E}_{x \in V} e(\phi_{h_1}(x) + \phi_{h_2}(x+h_1-h_4) - \phi_{h_3}(x) - \phi_{h_4}(x+h_1-h_4))| \gg 1.$$

In particular, by the equidistribution theory in Section 1.4, the polynomial $\phi_{h_1} + \phi_{h_2} - \phi_{h_3} - \phi_{h_4}$ is low rank.

The above discussion is valid in any value of $d \geq 2$, but is particularly simple when $d = 2$, as the ϕ_h are now linear, and so $\phi_{h_1} + \phi_{h_2} - \phi_{h_3} - \phi_{h_4}$ is now *constant*. Writing $\phi_h(x) = \xi_h \cdot x + \theta_h$ for some $\xi_h \in V$ using the standard dot product on V, and some (irrelevant) constant term $\theta_h \in \mathbf{F}$, we conclude that

(1.44) $$\xi_{h_1} + \xi_{h_2} = \xi_{h_3} + \xi_{h_4}$$

for many additive quadruples h_1, h_2, h_3, h_4.

We now have to solve an additive combinatorics problem, namely to classify the functions $h \mapsto \xi_h$ from V to V which are "1% affine linear" in the sense that the property (1.44) holds for many additive quadruples; equivalently, the graph $\{(h, \xi_h) : h \in H\}$ in $V \times V$ has high "additive energy", defined as the number of additive quadruples that it contains. An obvious example of a function with this property is an affine-linear function $\xi_h = Mh + \xi_0$, where $M \colon V \to V$ is a linear transformation and $\xi_0 \in V$. As it turns out, this is essentially the only example:

Proposition 1.5.8 (Balog-Szemerédi-Gowers-Freiman theorem for vector spaces). *Let $H \subset V$, and let $h \mapsto \xi_h$ be a map from H to V such that (1.44) holds for $\gg |V|^3$ additive quadruples in H. Then there exists an affine function $h \mapsto Mh + \xi_0$ such that $\xi_h = Mh + \xi_0$ for $\gg |V|$ values of h in H.*

This proposition is a consequence of standard results in additive combinatorics, in particular, the Balog-Szemerédi-Gowers lemma and Freiman's

1.5. Inverse conjecture over finite fields

theorem for vector spaces; see [**TaVu2006**, §11.3] for further discussion. The proof is elementary but a little lengthy and would take us too far afield, so we simply assume this proposition for now and keep going. We conclude that

(1.45) $$|\mathbf{E}_{x \in V} \Delta_h f(x) e(Mh \cdot x) e(\xi_0 \cdot x)| \gg 1$$

for many $h \in V$.

The most difficult term to deal with here is the quadratic term $Mh \cdot x$. To deal with this term, suppose temporarily that M is symmetric, thus $Mh \cdot x = Mx \cdot h$. Then (since we are in odd characteristic) we can *integrate* $Mh \cdot x$ as

$$Mh \cdot x = \partial_h \left(\frac{1}{2} Mx \cdot x\right) - \frac{1}{2} Mh \cdot h$$

and thus

$$|\mathbf{E}_{x \in V} f(x+h) e(\tfrac{1}{2} M(x+h) \cdot (x+h)) \overline{f(x)} e(-\tfrac{1}{2} Mx \cdot x) e(\xi_0 \cdot x)| \gg 1$$

for many $h \in H$. Taking L^2 norms in h, we conclude that the U^2 inner product between two copies of $f(x) e(\tfrac{1}{2} Mx \cdot x)$ and two copies of $f(x) e(\tfrac{1}{2} Mx \cdot x) e(-\xi_0 \cdot x)$ is $\gg 1$. Applying the U^2 Cauchy-Schwarz-Gowers inequality, followed by the U^2 inverse theorem, we conclude that $f(x) e(\tfrac{1}{2} Mx \cdot x)$ correlates with $e(\phi)$ for some linear phase, and thus f itself correlates with $e(\psi)$ for some quadratic phase.

This argument also works (with minor modification) when M is *virtually symmetric*, in the sense that there exist a bounded index subspace of V such that the restriction of the form $Mh \cdot x$ to V is symmetric, by foliating into cosets of that subspace; we omit the details. On the other hand, if M is not virtually symmetric, there is no obvious way to "integrate" the phase $e(Mh \cdot x)$ to eliminate it as above. (Indeed, in order for $Mh \cdot x$ to be "exact" in the sense that it is the "derivative" of something (modulo lower order terms), e.g., $Mh \cdot x \approx \partial_h \Phi$ for some Φ, it must first be "closed" in the sense that $\partial_k (Mh \cdot x) \approx \partial_h (Mk \cdot x)$ in some sense, since we have $\partial_h \partial_k = \partial_k \partial_h$; thus we again see the emergence of cohomological concepts in the background.)

To establish the required symmetry on M, we return to Gowers' argument from Lemma 1.5.7, and tweak it slightly. We start with (1.45) and rewrite it as

$$|\mathbf{E}_{x \in V} f(x+h) f'(x) e(Mh \cdot x)| \gg 1$$

where $f'(x) := \overline{f(x)} e(\xi_0 \cdot x)$. We square-average this in h to obtain

$$|\mathbf{E}_{x,y,h \in V} f(x+h) f'(x) \overline{f(y+h)} \overline{f'(y)} e(Mh \cdot (x-y))| \gg 1.$$

Now we make the somewhat unusual substitution $z = x + y + h$ to obtain

$$|\mathbf{E}_{x,y,z \in V} f(z-y) f'(x) \overline{f(z-x)} \overline{f'(y)} e(M(z-x-y) \cdot (x-y))| \gg 1.$$

Thus there exists z such that
$$|\mathbf{E}_{x,y \in V} f(z-y) f'(x) \overline{f(z-x) f'(y)} e(M(z-x-y) \cdot (x-y))| \gg 1.$$
We collect all terms that depend only on x (and z) or only on y (and z) to obtain
$$|\mathbf{E}_{x,y \in V} f_{z,1}(x) f_{z,2}(y) e(Mx \cdot y - My \cdot x)| \gg 1$$
for some bounded functions $f_{z,1}, f_{z,2}$. Eliminating these functions by two applications of Cauchy-Schwarz, we obtain
$$|\mathbf{E}_{x,y,x',y' \in V} e(M(x-x') \cdot (y-y') - M(y-y') \cdot (x-x'))| \gg 1$$
or, on making the change of variables $a := x - x', b := y - y'$,
$$|\mathbf{E}_{a,b \in V} e(Ma \cdot b - Mb \cdot a)| \gg 1.$$
Using equidistribution theory, this means that the quadratic form $(a,b) \mapsto Ma \cdot b - Mb \cdot a$ is low rank, which easily implies that M is virtually symmetric.

Remark 1.5.9. In [**Sa2007**] a variant of this argument was introduced to deal with the even characteristic case. The key new idea is to split the matrix of M into its diagonal component, plus the component that vanishes on the diagonal. The latter component can made (virtually) (anti-)symmetric and thus expressible as $U + U^T$ where U is an upper-triangular matrix; this allows for an integration as before, using $Ux \cdot x$ in place of $\frac{1}{2} Mx \cdot x$. In characteristic two, the diagonal contribution to $Mx \cdot x$ is linear in x and can be easily handled by passing to a codimension one subspace. See [**Sa2007**] for details.

Now we turn to the general d case. In principle, the above argument should still work, say for $d = 3$. The main sticking point is that instead of dealing with a vector-valued function $h \mapsto \xi_h$ that is approximately linear in the sense that (1.44) holds for many additive quadruples, in the $d = 3$ case one is now faced with a *matrix-valued* function $h \mapsto M_h$ with the property that
$$M_{h_1} + M_{h_2} = M_{h_3} + M_{h_4} + L_{h_1,h_2,h_3,h_4}$$
for many additive quadruples h_1, h_2, h_3, h_4, where the matrix L_{h_1,h_2,h_3,h_4} has bounded rank. With our current level of additive combinatorics technology, we are not able to deal properly with this bounded rank error (the main difficulty being that the set of low rank matrices has no good "doubling" properties). Because of this obstruction, no generalisation of the above arguments to higher d has been found.

There is, however, another approach, based ultimately on the ergodic theory work of Host and Kra [**HoKr2005**] and of Ziegler [**Zi2007**], that can handle the general d case, which was worked out in [**TaZi2010**] and [**BeTaZi2010**]. It turns out that it is convenient to phrase these arguments

1.5. Inverse conjecture over finite fields

in the language of ergodic theory. However, in order not to have to introduce too much additional material, we will describe the arguments here in the case $d = 3$ without explicitly using ergodic theory notation. To do this, though, we will have to sacrifice a lot of rigour and only work with some illustrative special cases rather than the general case, and also use somewhat vague terminology (e.g. "general position" or "low rank").

To simplify things further, we will establish the U^3 inverse theorem only for a special type of function, namely a quartic phase[14] $e(\phi)$, where $\phi\colon V \to \mathbf{F}$ is a classical polynomial of degree 4. The claim to show then is that if $\|e(\phi)\|_{U^3(V)} \gg 1$, then $e(\phi)$ correlates with a cubic phase. In the high characteristic case $p > 4$, this result can be handled by equidistribution theory. Indeed, since

$$\|e(\phi)\|_{U^3(V)}^8 = \mathbf{E}_{x,h_1,h_2,h_3,h_4} e(\partial_{h_1}\partial_{h_2}\partial_{h_3}\partial_{h_4}\phi(x)),$$

that theory tells us that the quartic polynomial

$$(x, h_1, h_2, h_3, h_4) \mapsto \partial_{h_1}\partial_{h_2}\partial_{h_3}\partial_{h_4}\phi(x)$$

is low rank. On the other hand, in high characteristic one has the Taylor expansion

$$\phi(x) = \frac{1}{4!}\partial_x\partial_x\partial_x\partial_x\phi(0) + Q(x)$$

for some cubic function Q (as can be seen for instance by decomposing into monomials). From this we easily conclude that ϕ itself has low rank (i.e. it is a function of boundedly many cubic (or lower degree) polynomials), at which point it is easy to see from Fourier analysis that $e(\phi)$ will correlate with the exponential of a polynomial of degree at most 3.

Now we present a different argument that relies slightly less on the quartic nature of ϕ; it is a substantially more difficult argument, and we will skip some steps here to simplify the exposition, but the argument happens to extend to more general situations. As $\|e(\phi)\|_{U^3} \gg 1$, we have $\|\Delta_h e(\phi)\|_{U^2} \gg 1$ for many h, thus by the inverse U^2 theorem, $\Delta_h e(\phi) = e(\partial_h\phi)$ correlates with a quadratic phase. Using equidistribution theory, we conclude that the cubic polynomial $\partial_h\phi$ is low rank.

At present, the low rank property for $\partial_h\phi$ is only true for many h. But from the cocycle identity

(1.46) $$\partial_{h+k}\phi = \partial_h\phi + T^h\partial_k\phi,$$

we see that if $\partial_h\phi$ and $\partial_k\phi$ are both low rank, then so is $\partial_{h+k}\phi$; thus the property of $\partial_h\phi$ being low rank is in some sense preserved by addition. Using this and a bit of additive combinatorics, one can conclude that $\partial_h\phi$ is low

---

[14] A good example to keep in mind is the symmetric polynomial phase $e(S_2/2)$ from Section 1.5.2, though one has to take some care with this example due to the low characteristic.

rank for all h in a bounded index subspace of V; restricting to that subspace, we will now assume that $\partial_h \phi$ is low rank for *all* $h \in V$. Thus we have

$$\partial_h \phi = F_h(\vec{Q}_h)$$

where $\vec{Q}_h$ is some bounded collection of quadratic polynomials for each h, and F_h is some function. To simplify the discussion, let us pretend that $\vec{Q}_h$ in fact consists of just a single quadratic Q_h, plus some linear polynomials $\vec{L}_h$, thus

(1.47) $$\partial_h \phi = F_h(Q_h, \vec{L}_h).$$

There are two extreme cases to consider, depending on how Q_h depends on h. Consider first a "core" case when $Q_h = Q$ is independent of h. Thus

(1.48) $$\partial_h \phi = F_h(Q, \vec{L}_h).$$

If Q is low rank, then we can absorb it into the L_h factors, so suppose instead that Q is high rank, and thus equidistributed even after fixing the values of L_h.

The function $\partial_h \phi$ is cubic, and Q is a high rank quadratic. Because of this, the function $F'(Q, L_h)$ must be at most linear in the Q variable; this can be established by another application of equidistribution theory; see [**GrTa2009**, §8]. Thus one can factorise

$$\partial_h \phi = Q F'_h(L_h) + F''_h(L_h)$$

for some functions F'_h, F''_h. In fact, as $\partial_h \phi$ is cubic, F'_h must be linear, while F''_h is cubic.

By comparing the Q coefficients $F'_h(L_h)$ in the cocycle equation (1.46), we see that the function $\rho_h := F'_h(L_h)$ is itself a cocycle:

$$\rho_{h+k} = \rho_h + T^h \rho_k.$$

As a consequence, we have $\rho_h = \partial_h R$ for some function $R: V \to \mathbf{R}/\mathbf{Z}$. Since ρ_h is linear, R is quadratic; thus we have

(1.49) $$\partial_h \phi = Q \partial_h R + F''_h(L_h).$$

With a high characteristic assumption $p > 2$, one can ensure R is classical. We will assume that R is high rank, as this is the most difficult case.

Suppose first that $Q = R$. In high characteristic, one can then integrate $Q \partial_h Q$ by expressing this as $\partial_h(\frac{1}{2}Q^2)$ plus lower order terms, thus $\partial_h(\phi - \frac{1}{2}Q^2)$ is an order 1 function in the sense that it is a function of a bounded number of linear functions. In particular, $e(\partial_h(\phi - \frac{1}{2}Q^2))$ has a large U^2 norm for all h, which implies that $e(\phi - \frac{1}{2}Q^2)$ has a large U^3 norm, and thus correlates with a quadratic phase. Since $e(\frac{1}{2}Q^2)$ can be decomposed by Fourier analysis into a linear combination of quadratic phases, we conclude that $e(\phi)$ correlates with a quadratic phase and one is thus done in this case.

1.5. Inverse conjecture over finite fields

Now consider the other extreme, in which Q and R lie in general position. Then, if we differentiate (1.49) in k, we obtain that one has

$$\partial_k \partial_h \phi = \partial_k Q \partial_h R + Q \partial_k \partial_h R + \partial_k Q (\partial_k \partial_h R) + \partial_k F''_h(L_h),$$

and then anti-symmetrising in k, h one has

$$0 = \partial_k Q \partial_h R - \partial_h Q \partial_k R + (\partial_k Q - \partial_h Q) \partial_k \partial_h R + \partial_k F''_h(L_h) - \partial_h F''_k(L_h).$$

If Q and R are unrelated, then the linear forms $\partial_k Q, \partial_k R$ will typically be in general position with respect to each other and with L_h, and similarly $\partial_h Q, \partial_h R$ will be in general position with respect to each other and with L_k. From this, one can show that the above equation is not satisfiable generically, because the mixed terms $\partial_k Q \partial_h R - \partial_h Q \partial_k R$ cannot be cancelled by the simpler terms in the above expression.

An interpolation of the above two arguments can handle the case in which Q_h does not depend on h. Now we consider the other extreme, in which Q_h varies in h, so that Q_h and Q_k are in general position for generic h, k, and similarly[15] for Q_h and Q_{h+k}, or for Q_k and Q_{h+k}.

To analyse this situation, we return to the cocycle equation (1.46), which currently reads

(1.50) $$F_{h+k}(Q_{h+k}, \vec{L}_{h+k}) = F_h(Q_h, \vec{L}_h) + T^h F_k(Q_k, \vec{L}_k).$$

Because any two of Q_{h+k}, Q_h, Q_k can be assumed to be in general position, one can show using equidistribution theory that the above equation can only be satisfied when the F_h are linear in the Q_h variable, thus

$$\partial_h \phi = Q_h F'_h(\vec{L}_h) + F''_h(\vec{L}_h)$$

much as before. Furthermore, the coefficients $F'_h(\vec{L}_h)$ must now be (essentially) constant in h in order to obtain (1.50). Absorbing this constant into the definition of Q_h, we now have

$$\partial_h \phi = Q_h + F''_h(\vec{L}_h).$$

We will once again pretend that $\vec{L}_h$ is just a single linear form L_h. Again we consider two extremes. If $L_h = L$ is independent of h, then by passing to a bounded index subspace (the level set of L) we now see that $\partial_h \phi$ is quadratic, hence ϕ is cubic, and we are done. Now suppose instead that L_h varies in h, so that L_h, L_k are in general position for generic h, k. We look at the cocycle equation again, which now tells us that $F''_h(\vec{L}_h)$ obeys the *quasicocycle* condition

$$Q_{h,k} + F''_{h+k}(\vec{L}_{h+k}) = F''_h(\vec{L}_h) + T^h F''_k(\vec{L}_k)$$

---

[15] Note though that we cannot simultaneously assume that Q_h, Q_k, Q_{h+k} are in general position; indeed, Q_h might vary linearly in h, and indeed we expect this to be the basic behaviour of Q_h here, as was observed in the preceding argument.

where $Q_{h,k} := Q_{h+k} - Q_h - T^h Q_k$ is a quadratic polynomial. With any two of L_h, L_k, L_{h+k} in general position, one can then conclude (using equidistribution theory) that F_h'', F_k'', F_{h+k}'' are quadratic polynomials. Thus $\partial_h \phi$ is quadratic, and ϕ is cubic as before. This completes the heuristic discussion of various extreme model cases; the general case is handled by a rather complicated combination of all of these special case methods, and is best performed[16] in the framework of ergodic theory; see [**BeTaZi2010**]. The various functional equations for these vertical derivatives were first introduced by Conze and Lesigne [**CoLe1984**].

1.5.4. Consequences of the inverse conjecture for the Gowers norm.
We now discuss briefly some of the consequences of the inverse conjecture for the Gowers norm, beginning with Szemerédi's theorem in vector fields (Theorem 1.5.4). We will use the density increment method[17]. Let $A \subset V = \mathbf{F}^n$ be a set of density at least δ containing no lines. This implies that the p-linear form

$$\Lambda(1_A, \ldots, 1_A) := \mathbf{E}_{x,r \in \mathbf{F}^n} 1_A(x) \ldots 1_A(x + (p-1)r)$$

has size $o(1)$. On the other hand, as this pattern has complexity $p - 2$, we see from Section 1.3 that one has the bound

$$|\Lambda(f_0, \ldots, f_{p-1})| \leq \sup_{0 \leq j \leq p-1} \|f_j\|_{U^{p-1}(V)}$$

whenever $f_0, \ldots, f_{p-1}$ are bounded in magnitude by 1. Splitting $1_A = \delta + (1_A - \delta)$, we conclude that

$$\Lambda(1_A, \ldots, 1_A) = \delta^p + O_p(\|1_A - \delta\|_{U^{p-1}(V)})$$

and thus (for n large enough)

$$\|1_A - \delta\|_{U^{p-1}(V)} \gg_{p,\delta} 1.$$

Applying Theorem 1.5.3, we find that there exists a polynomial ϕ of degree at most $p - 2$ such that

$$|\langle 1_A - \delta, e(\phi) \rangle| \gg_{p,\delta} 1.$$

To proceed we need the following analogue of Proposition 1.2.6:

Exercise 1.5.6 (Fragmenting a polynomial into subspaces). Let $\phi \colon \mathbf{F}^n \to \mathbf{F}$ be a classical polynomial of degree $d < p$. Show that one can partition V into affine subspaces W of dimension at least $n'(n, d, p)$, where $n' \to \infty$ as $n \to \infty$ for fixed d, p, such that ϕ is constant on each W. (*Hint*: Induct

---

[16] In particular, the idea of extracting out the coefficient of a key polynomial, such as the coefficient $F_h'(L_h)$ of Q, is best captured by the ergodic theory concept of *vertical differentiation*. Again, see [**BeTaZi2010**] for details.

[17] An energy increment argument is also possible, but is more complicated; see [**GrTa2010b**].

1.5. Inverse conjecture over finite fields

on d, and use Exercise 1.4.6 repeatedly to find a good initial partition into subspaces on which ϕ has degree at most $d - 1$.)

Exercise 1.5.7. Use the previous exercise to complete the proof of Theorem 1.5.4. (*Hint:* Mimic the density increment argument from Section 1.2.)

By using the inverse theorem as a substitute for Lemma 1.2.8, one obtains the following regularity lemma, analogous to Theorem 1.2.11:

Theorem 1.5.10 (Strong arithmetic regularity lemma). *Suppose that* $\operatorname{char}(\mathbf{F}) = p > d \geq 0$. *Let* $f\colon V \to [0,1]$, *let* $\varepsilon > 0$, *and let* $F\colon \mathbf{R}^+ \to \mathbf{R}^+$ *be an arbitrary function. Then we can decompose* $f = f_{\mathrm{str}} + f_{\mathrm{sml}} + f_{\mathrm{psd}}$ *and find* $1 \leq M = O_{\varepsilon, F, d, p}(1)$ *such that*

   (i) (*Nonnegativity*) $f_{\mathrm{str}}, f_{\mathrm{str}} + f_{\mathrm{sml}}$ *take values in* $[0, 1]$, *and* $f_{\mathrm{sml}}, f_{\mathrm{psd}}$ *have mean zero;*

   (ii) (*Structure*) f_{str} *is a function of* M *classical polynomials of degree at most* d;

   (iii) (*Smallness*) f_{sml} *has an* $L^2(V)$ *norm of at most* ε; *and*

   (iv) (*Pseudorandomness*) *One has* $\|f_{\mathrm{psd}}\|_{U^{d+1}(V)} \leq 1/F(M)$ *for all* $\alpha \in \mathbf{R}$.

For a proof, see [**Ta2007**]. The argument is similar to that appearing in Theorem 1.2.11, but the discrete nature of polynomials in bounded characteristic allows one to avoid a number of technical issues regarding measurability.

This theorem can then be used for a variety of applications in additive combinatorics. For instance, it gives the following variant of a result of Bergelson, Host, and Kra [**BeHoKa2005**]:

Proposition 1.5.11. *Let* $p > 4 \geq k$, *let* $\mathbf{F} = \mathbf{F}_p$, *and let* $A \subset \mathbf{F}^n$ *with* $|A| \geq \delta |\mathbf{F}^n|$, *and let* $\varepsilon > 0$. *Then for* $\gg_{\delta, \varepsilon, p} |\mathbf{F}^n|$ *values of* $h \in \mathbf{F}^n$, *one has*

$$|\{x \in \mathbf{F}^n : x, x + h, \ldots, x + (k-1)h \in A\}| \geq (\delta^k - \varepsilon)|\mathbf{F}^n|.$$

Roughly speaking, the idea is to apply the regularity lemma to $f := 1_A$, discard the contribution of the f_{sml} and f_{psd} errors, and then control the structured component using the equidistribution theory from Section 1.4. A proof of this result can be found in [**Gr2007**]; see also [**GrTa2010b**] for an analogous result in $\mathbf{Z}/N\mathbf{Z}$. Curiously, the claim fails when 4 is replaced by any larger number; this is essentially an observation of Ruzsa that appears in the appendix of [**BeHoKa2005**].

The above regularity lemma (or more precisely, a close relative of this lemma) was also used in [**GoWo2010b**]:

Theorem 1.5.12 (Gowers-Wolf theorem [**GoWo2010b**]). *Let $\Psi = (\psi_1, \ldots, \psi_t)$ be a collection of linear forms with integer coefficients, with no two forms being linearly dependent. Let $\mathbf{F}$ have sufficiently large characteristic, and suppose that $f_1, \ldots, f_t \colon \mathbf{F}^n \to \mathbf{C}$ are functions bounded in magnitude by 1 such that*

$$|\Lambda_\Psi(f_1, \ldots, f_t)| \geq \delta$$

where Λ_Ψ was the form defined in Section 1.3. Then for each $1 \leq i \leq t$ there exists a classical polynomial ϕ_i of degree at most d such that

$$|\langle f_i, e(\phi_i)\rangle_{L^2(\mathbf{F}^n)}| \gg_{d,\Psi,\delta} 1,$$

where d is the true complexity of the system Ψ as defined in Section 1.3. This d is best possible.

1.6. The inverse conjecture for the Gowers norm II. The integer case

In Section 1.5, we saw that the Gowers uniformity norms on vector spaces $\mathbf{F}^n$ were controlled by classical polynomial phases $e(\phi)$.

Now we study the analogous situation on cyclic groups $\mathbf{Z}/N\mathbf{Z}$. Here, there is an unexpected surprise: the polynomial phases (classical or otherwise) are no longer sufficient to control the Gowers norms $U^{s+1}(\mathbf{Z}/N\mathbf{Z})$ once s exceeds 1. To resolve this problem, one must enlarge the space of polynomials to a larger class. It turns out that there are at least three closely related options for this class: the *local polynomials*, the *bracket polynomials*, and the *nilsequences*. Each of the three classes has its own strengths and weaknesses, but in my opinion the nilsequences seem to be the most natural class, due to the rich algebraic and dynamical structure coming from the nilpotent Lie group undergirding such sequences. For reasons of space we shall focus primarily on the nilsequence viewpoint here.

Traditionally, nilsequences have been defined in terms of linear orbits $n \mapsto g^n x$ on nilmanifolds G/Γ; however, in recent years it has been realised that it is convenient for technical reasons (particularly for the quantitative "single-scale" theory) to generalise this setup to that of *polynomial* orbits $n \mapsto g(n)\Gamma$, and this is the perspective we will take here.

A polynomial phase $n \mapsto e(\phi(n))$ on a finite abelian group H is formed by starting with a polynomial $\phi \colon H \to \mathbf{R}/\mathbf{Z}$ to the unit circle, and then composing it with the exponential function $e \colon \mathbf{R}/\mathbf{Z} \to \mathbf{C}$. To create a nilsequence $n \mapsto F(g(n)\Gamma)$, we generalise this construction by starting with a polynomial $g\Gamma\colon H \to G/\Gamma$ into a *nilmanifold* G/Γ, and then composing this with a Lipschitz[18] function $F\colon G/\Gamma \to \mathbf{C}$. These classes of sequences certainly include the polynomial phases, but are somewhat more general; for instance, they *almost*[19] include *bracket polynomial* phases such as $n \mapsto e(\lfloor \alpha n \rfloor \beta n)$.

In this section we set out the basic theory for these nilsequences, including their equidistribution theory (which generalises the equidistribution theory of polynomial flows on tori from Section 1.1) and show that they are indeed obstructions to the Gowers norm being small. This leads to the *inverse conjecture for the Gowers norms* that shows that the Gowers norms on cyclic groups are indeed controlled by these sequences.

1.6.1. General theory of polynomial maps. In previous sections, we defined the notion of a (non-classical) polynomial map ϕ of degree at most d between two additive groups H, G, to be a map $\phi\colon H \to G$ obeying the identity
$$\partial_{h_1} \ldots \partial_{h_{d+1}} \phi(x) = 0$$
for all $x, h_1, \ldots, h_{d+1} \in H$, where $\partial_h \phi(x) := \phi(x+h) - \phi(x)$ is the additive discrete derivative operator.

There is another way to view this concept. For any $k, d \geq 0$, define the *Host-Kra group* $\mathrm{HK}^k(H, \leq d)$ of H of dimension k and degree d to be the subgroup of $H^{\{0,1\}^d}$ consisting of all tuples $(x_\omega)_{\omega \in \{0,1\}^k}$ obeying the constraints
$$\sum_{\omega \in F} (-1)^{|\omega|} x_\omega = 0$$
for all faces F of the unit cube $\{0,1\}^k$ of dimension at least $d+1$, where $|(\omega_1, \ldots, \omega_k)| := \omega_1 + \cdots + \omega_k$. (These constraints are of course trivial if $k \leq d$.) An r-dimensional face of the unit cube $\{0,1\}^k$ is of course formed by freezing $k-r$ of the coordinates to a fixed value in $\{0,1\}$, and letting the remaining r coordinates vary freely in $\{0,1\}$.

Thus, for instance, $\mathrm{HK}^2(H, \leq 1)$ is (essentially) the space of parallelograms $(x, x+h, x+k, x+h+k)$ in H^4, while $\mathrm{HK}^2(H, \leq 0)$ is the diagonal group $\{(x,x,x,x) : x \in H^4\}$, and $\mathrm{HK}^2(H, \leq 2)$ is all of H^4.

Exercise 1.6.1. Let $\phi\colon H \to G$ be a map between additive groups, and let $k > d \geq 0$. Show that ϕ is a (non-classical) polynomial of degree at most d if it maps $\mathrm{HK}^k(H, \leq 1)$ to $\mathrm{HK}^k(G, \leq d)$, i.e., that $(\phi(x_\omega))_{\omega \in \{0,1\}^k} \in \mathrm{HK}^k(G, \leq d)$ whenever $(x_\omega)_{\omega \in \{0,1\}^k} \in \mathrm{HK}^k(H, \leq 1)$.

---

[18] The Lipschitz regularity class is convenient for minor technical reasons, but one could also use other regularity classes here if desired.

[19] The "almost" here is because the relevant functions $F\colon G/\Gamma \to \mathbf{C}$ involved are only piecewise Lipschitz rather than Lipschitz, but this is primarily a technical issue and one should view bracket polynomial phases as "morally" being nilsequences.

It turns out (somewhat remarkably) that these notions can be satisfactorily generalised to a non-abelian setting, this was first observed by Leibman [**Le1998, Le2002**]. The (now multiplicative) groups H, G need to be equipped with an additional structure, namely that of a *filtration*.

Definition 1.6.1 (Filtration). A *filtration* on a multiplicative group G is a family $(G_{\geq i})_{i=0}^\infty$ of subgroups of G obeying the nesting property
$$G \geq G_{\geq 0} \geq G_{\geq 1} \geq \ldots$$
and the filtration property
$$[G_{\geq i}, G_{\geq j}] \subset G_{\geq i+j}$$
for all $i, j \geq 0$, where $[H, K]$ is the group generated by $\{[h, k] : h \in H, k \in K\}$, where $[h, k] := hkh^{-1}k^{-1}$ is the commutator of h and k. We will refer to the pair $G_\bullet = (G, (G_{\geq i})_{i=0}^\infty)$ as a *filtered group*. We say that an element g of G has *degree* $\geq i$ if it belongs to $G_{\geq i}$; thus, for instance, a degree $\geq i$ and degree $\geq j$ element will commute modulo $\geq i + j$ errors.

In practice we usually have $G_{\geq 0} = G$. As such, we see that $[G, G_{\geq j}] \subset G_{\geq j}$ for all j, and so all the $G_{\geq j}$ are normal subgroups of G.

Exercise 1.6.2. Define the *lower central series*
$$G = G_0 = G_1 \geq G_2 \geq \ldots$$
of a group G by setting $G_0, G_1 := G$ and $G_{i+1} := [G, G_i]$ for $i \geq 1$. Show that the lower central series $(G_j)_{j=0}^\infty$ is a filtration of G. Furthermore, show that the lower central series is the minimal filtration that starts at G, in the sense that if $(G'_{\geq j})_{j=0}^\infty$ is any other filtration with $G'_{\geq 0} = G$, then $G'_{\geq j} \supset G_{\geq j}$ for all j.

Example 1.6.2. If G is an abelian group, and $d \geq 0$, we define the *degree d filtration* $(G, \leq d)$ on G by setting $G_{\geq i} := G$ if $i \leq d$ and $G_{\geq i} = \{\mathrm{id}\}$ for $i > d$.

Example 1.6.3. If $G_\bullet = (G, (G_{\geq i})_{i=0}^\infty)$ is a filtered group, and $k \geq 0$, we define the shifted filtered group $G_\bullet^{+k} := (G, (G_{\geq i+k})_{i=0}^\infty)$; this is clearly again a filtered group.

Definition 1.6.4 (Host-Kra groups). Let $G_\bullet = (G, (G_{\geq i})_{i=0}^\infty)$ be a filtered group, and let $k \geq 0$ be an integer. The *Host-Kra group* $\mathrm{HK}^k(G_\bullet)$ is the subgroup of $G^{\{0,1\}^k}$ generated by the elements g_F with F an arbitrary face in $\{0, 1\}^k$ and g an element of $G_{\geq k-\dim(F)}$, where g_F is the element of $G^{\{0,1\}^k}$ whose coordinate at ω is equal to g when $\omega \in F$ and equal to $\{\mathrm{id}\}$ otherwise.

From construction we see that the Host-Kra group is symmetric with respect to the symmetry group $S_k \ltimes (\mathbf{Z}/2\mathbf{Z})^k$ of the unit cube $\{0, 1\}^k$. We will use these symmetries implicitly in the sequel without further comment.

1.6. Inverse conjecture over the integers

Example 1.6.5. Let us parameterise an element of $G^{\{0,1\}^2}$ as $(g_{00}, g_{01}, g_{10}, g_{11})$. Then $\mathrm{HK}^2(G)$ is generated by elements of the form (g_0, g_0, g_0, g_0) for $g_0 \in G_{\geq 0}$, $(\mathrm{id}, \mathrm{id}, g_1, g_1)$ and $(\mathrm{id}, g_1, \mathrm{id}, g_1)$, and $(\mathrm{id}, \mathrm{id}, \mathrm{id}, g_2)$ for $g_0 \in G_{\geq 0}, g_1 \in G_{\geq 1}, g_2 \in G_{\geq 2}$. (This does not cover all the possible faces of $\{0,1\}^2$, but it is easy to see that the remaining faces are redundant.) In other words, $\mathrm{HK}^2(G)$ consists of all group elements of the form $(g_0, g_0 g_1, g_0 g_1', g_0 g_1 g_1' g_2)$, where $g_0 \in G_{\geq 0}$, $g_1, g_1' \in G_{\geq 1}$, and $g_2 \in G_{\geq 2}$. This example is generalised in the exercise below.

Exercise 1.6.3. Define a *lower face* to be a face of a discrete cube $\{0,1\}^k$ in which all the frozen coefficients are equal to 0. Let us order the lower faces as $F_1, \ldots, F_{2^k-1}$ in such a way that $i \geq j$ whenever F_i is a subface of F_j. Let $G_\bullet$ be a filtered group. Show that every element of $\mathrm{HK}^k(G_\bullet)$ has a unique representation of the form $\prod_{i=0}^{2^k-1} (g_i)_{F_i}$, where $g_i \in G_{\geq k-\dim(F_i)}$ and the product is taken from left to right (say).

Exercise 1.6.4. If G is an abelian group, show that the group $\mathrm{HK}^k(G, \leq d)$ defined in Definition 1.6.4 agrees with the group defined at the beginning of this section for additive groups (after transcribing the former to multiplicative notation).

Exercise 1.6.5. Let $G_\bullet$ be a filtered group. Let F be an r-dimensional face of $\{0,1\}^k$. Identifying F with $\{0,1\}^r$ in an obvious manner, we then obtain a restriction homomorphism from $G^{\{0,1\}^k}$ with $G^F \equiv G^{\{0,1\}^r}$. Show that the restriction of any element of $\mathrm{HK}^k(G_\bullet)$ to $G^{\{0,1\}^r}$ then lies in $\mathrm{HK}^r(G_\bullet)$.

Exercise 1.6.6. Let $G_\bullet$ be a filtered group, let $k \geq 0$ and $l \geq 1$ be integers, and let $g = (g_\omega)_{\omega \in \{0,1\}^k}$ and $h = (h_\omega)_{\omega \in \{0,1\}^k}$ be elements of $G^{\{0,1\}^k}$. Let $f = (f_\omega)_{\omega \in \{0,1\}^{k+l}}$ be the element of $G^{\{0,1\}^{k+l}}$ defined by setting f_{ω_k, ω_l} for $\omega_k \in \{0,1\}^k, \omega_l \in \{0,1\}^l$ to equal g_{ω_k} for $\omega_l \neq (1, \ldots, 1)$, and equal to $g_{\omega_k} h_{\omega_k}$ otherwise. Show that $f \in \mathrm{HK}^{k+l}(G_\bullet)$ if and only if $g \in \mathrm{HK}^k(G_\bullet)$ and $h \in \mathrm{HK}^k(G_\bullet^{+l})$, where $G_\bullet^{+l}$ is defined in Example 1.6.3. (*Hint:* Use Exercises 1.6.3, 1.6.5.)

Exercise 1.6.7. Let $G_\bullet$ be a filtered group, let $k \geq 1$, and let $g = (g_\omega)_{\omega \in \{0,1\}^k}$ be an element of $G^{\{0,1\}^k}$. We define the *derivative* $\partial_1 g \in G^{\{0,1\}^{k-1}}$ in the first variable to be the tuple $(g_{\omega,1} g_{\omega,0}^{-1})_{\omega \in \{0,1\}^{k-1}}$. Show that $g \in \mathrm{HK}^k(G_\bullet)$ if and only if the restriction of g to $\{0,1\}^{k-1}$ lies in $\mathrm{HK}^{k-1}(G_\bullet)$ and $\partial_1 g$ lies in $\mathrm{HK}^k(G_\bullet^{+1})$, where $G_\bullet^{+1}$ is defined in Example 1.6.3.

Remark 1.6.6. The Host-Kra groups of a filtered group in fact form a *cubic complex*, a concept used in topology; but we will not pursue this connection here.

In analogy with Exercise 1.6.1, we can now define the general notion of a polynomial map:

Definition 1.6.7. A map $\phi\colon H \to G$ between two filtered groups $H_\bullet, G_\bullet$ is said to be *polynomial* if it maps $\mathrm{HK}^k(H_\bullet)$ to $\mathrm{HK}^k(G_\bullet)$ for each $k \geq 0$. The space of all such maps is denoted $\mathrm{Poly}(H_\bullet \to G_\bullet)$.

Since $\mathrm{HK}^k(H_\bullet), \mathrm{HK}^k(G_\bullet)$ are groups, we immediately obtain[20]

Theorem 1.6.8 (Lazard-Leibman theorem). $\mathrm{Poly}(H_\bullet \to G_\bullet)$ *forms a group under pointwise multiplication.*

In a similar spirit, we have

Theorem 1.6.9 (Filtered groups and polynomial maps form a category). *If $\phi\colon H \to G$ and $\psi\colon G \to K$ are polynomial maps between filtered groups $H_\bullet, G_\bullet, K_\bullet$, then $\psi \circ \phi\colon H \to K$ is also a polynomial map.*

We can also give some basic examples of polynomial maps. Any constant map from H to G taking values in $G_{\geq 0}$ is polynomial, as is any map $\phi\colon H \to G$ which is a *filtered homomorphism* in the sense that it is a homomorphism from $H_{\geq i}$ to $G_{\geq i}$ for any $i \geq 0$.

Now we turn to an alternate definition of a polynomial map. For any $h \in H$ and any map $\phi\colon H \to G$ Define the *multiplicative derivative* $\Delta_h \phi\colon H \to G$ by the formula $\Delta_h \phi(x) := \phi(hx)\phi(x)^{-1}$.

Theorem 1.6.10 (Alternate description of polynomials). *Let $\phi\colon H \to G$ be a map between two filtered groups H, G. Then ϕ is polynomial if and only if, for any $i_1, \ldots, i_m \geq 0$, $x \in H_{\geq 0}$, and $h_j \in H_{\geq i_j}$ for $j = 1, \ldots, m$, one has $\Delta_{h_1} \ldots \Delta_{h_m} \phi(x) \in G_{\geq i_1 + \cdots + i_m}$.*

In particular, from Exercise 1.6.1, we see that a non-classical polynomial of degree d from one additive group H to another G is the same thing as a polynomial map from $(H, \leq 1)$ to $(G, \leq d)$. More generally, a ϕ map from $(H, \leq 1)$ to a filtered group $G_\bullet$ is polynomial if and only if
$$\Delta_{h_1} \ldots \Delta_{h_i} \phi(x) \in G_{\geq i}$$
for all $i \geq 0$ and $x, h_1, \ldots, h_i \in H$.

Proof. We first prove the "only if" direction. It is clear (by using 0-dimensional cubes) that a polynomial map must map $H_{\geq 0}$ to $G_{\geq 0}$. To obtain the remaining cases, it suffices by induction on m to show that if ϕ

---

[20] From our choice of definitions, this theorem is a triviality, but the theorem is less trivial when using an alternate but non-trivially equivalent definition of a polynomial, which we will give shortly. Lazard [**La1954**] gave a version of this theorem when H was the integers and G was a nilpotent Lie group; the general problem of multiplying polynomial sequences was considered by Leibman [**Le1998**].

1.6. Inverse conjecture over the integers 97

is polynomial from $H_\bullet$ to $G_\bullet$, and $h \in H_{\geq i}$ for some $i \geq 0$, then $\Delta_h \phi$ is polynomial from $H_\bullet$ to $G_\bullet^{+i}$. But this is easily seen from Exercise 1.6.7.

Now we establish the "if" direction. We need to show that ϕ maps $\mathrm{HK}^k(H_\bullet)$ to $\mathrm{HK}^k(G_\bullet)$ for each k. We establish this by induction on k. The case $k = 0$ is trivial, so suppose that $k \geq 1$ and that the claim has already been established for all smaller values of k.

Let $h \in \mathrm{HK}^k(H_\bullet)$. We split $H^{\{0,1\}^k}$ as $H^{\{0,1\}^{k-1}} \times H^{\{0,1\}^{k-1}}$. From Exercise 1.6.7 we see that we can write $h = (h_0, h_1 h_0)$ where $h_0 \in \mathrm{HK}^{k-1}(H_\bullet)$ and $h_1 \in \mathrm{HK}^{k-1}(H_\bullet^{+1})$, thus $\phi(h) = (\phi(h_0), \phi(h_1 h_0))$ (extending ϕ to act on $H^{\{0,1\}^{k-1}}$ or $H^{\{0,1\}^k}$ in the obvious manner). By induction hypothesis, $\phi(h_0) \in \mathrm{HK}^{k-1}(G_\bullet)$, so by Exercise 1.6.7, it suffices to show that $\phi(h_1 h_0) \phi^{-1}(h_0) \in \mathrm{HK}^{k-1}(G_\bullet^{+1})$.

By telescoping series, it suffices to establish this when $h_1 = h_F$ for some face F of some dimension r in $\{0,1\}^{k-1}$ and some $h \in H_{\geq k-r}$, as these elements generate $\mathrm{HK}^{k-1}(H_\bullet^{+1})$. But then $\phi(h_1 h_0) \phi^{-1}(h_0)$ vanishes outside of F and is equal to $\Delta_{h_1} \phi(h_0)$ on F, so by Exercise 1.6.6 it will suffice to show that $\Delta_{h_1} \phi(h_0') \in \mathrm{HK}^r(G_\bullet^{+k-r})$, where h_0' is h_0 restricted to F (which one then identifies with $\{0,1\}^r$). But by the induction hypothesis, $\Delta_{h_1} \phi$ maps $\mathrm{HK}^r(H_\bullet)$ to $\mathrm{HK}^r(H_\bullet^{+k-r})$, and the claim then follows from Exercise 1.6.5. □

Exercise 1.6.8. Let $i_1, \ldots, i_k \geq 0$ be integers. If $G_\bullet$ is a filtered group, define $\mathrm{HK}^{(i_1,\ldots,i_k)}(G_\bullet)$ to be the subgroup of $G^{\{0,1\}^k}$ generated by the elements g_F, where F ranges over all faces of $\{0,1\}^k$ and $g \in G_{\geq i_{j_1} + \cdots + i_{j_r}}$, where $1 \leq j_1 < \cdots < j_r \leq k$ are the coordinates of F that are frozen. This generalises the Host-Kra groups $\mathrm{HK}^k(G_\bullet)$, which correspond to the case $i_1 = \cdots = i_k = 1$. Show that if ϕ is a polynomial map from $H_\bullet$ to $G_\bullet$, then ϕ maps $\mathrm{HK}^{(i_1,\ldots,i_k)}(H_\bullet)$ to $\mathrm{HK}^{(i_1,\ldots,i_k)}(G_\bullet)$.

Exercise 1.6.9. Suppose that $\phi: H \to G$ is a non-classical polynomial of degree $\leq d$ from one additive group to another. Show that ϕ is a polynomial map from $(H, \leq m)$ to $(G, \leq dm)$ for every $m \geq 1$. Conclude, in particular, that the composition of a non-classical polynomial of degree $\leq d$ and a non-classical polynomial of degree $\leq d'$ is a non-classical polynomial of degree $\leq dd'$.

Exercise 1.6.10. Let $\phi_1: H \to G_1$, $\phi_2: H \to G_2$ be non-classical polynomials of degrees $\leq d_1, \leq d_2$, respectively, between additive groups H, G_1, G_2, and let $B: G_1 \times G_2 \to G$ be a bihomomorphism to another additive group (i.e. B is a homomorphism in each variable separately). Show that $B(\phi_1, \phi_2): H \to G$ is a non-classical polynomial of degree $\leq d_1 + d_2$.

1.6.2. Nilsequences. We now specialise the above theory of polynomial maps $\phi: H \to G$ to the case when H is just the integers $\mathbf{Z} = (\mathbf{Z}, \leq 1)$ (viewed additively) and G is a nilpotent group. Recall that a group G is *nilpotent* of step at most s if the $(s+1)^{\text{th}}$ group G_{s+1} in the lower central series vanishes; thus, for instance, a group is nilpotent of step at most 1 if and only if it is abelian. Analogously, let us call a filtered group $G_\bullet$ *nilpotent* of degree at most s if $G_{\geq s+1}$ vanishes. Note that if $G_{\geq 0} = G$ and $G_\bullet$ is nilpotent of degree at most s, then G is nilpotent of step at most s. On the other hand, the degree of a filtered group can exceed the step; for instance, given an additive group G and an integer $d \geq 1$, $(G, \leq d)$ has degree d and step 1. The step is the traditional measure of nilpotency for groups, but the degree seems to be a more suitable measure in the filtered group category.

We refer to sequences $g \colon \mathbf{Z} \to G$ which are polynomial maps from $(\mathbf{Z}, \leq 1)$ to $G_\bullet$ as *polynomial sequences* adapted to $G_\bullet$. The space of all such sequences is denoted $\mathrm{Poly}(\mathbf{Z} \to G)$; by the machinery of the previous section, this is a multiplicative group. These sequences can be described explicitly:

Exercise 1.6.11. Let $s \geq 0$ be an integer, and let $G_\bullet$ be a filtered group which is nilpotent of degree s. Show that a sequence $g \colon \mathbf{Z} \to G$ is a polynomial sequence if and only if one has

$$(1.51) \qquad g(n) = g_0 g_1^{\binom{n}{1}} g_2^{\binom{n}{2}} \ldots g_s^{\binom{n}{s}}$$

for all $n \in \mathbf{Z}$ and some $g_i \in G_{\geq i}$ for $i = 0, \ldots, s$, where $\binom{n}{i} := \frac{n(n-1)\ldots(n-i+1)}{i!}$. Furthermore, show that the g_i are unique. We refer to the $g_0, \ldots, g_s$ as the *Taylor coefficients* of g at the origin.

Exercise 1.6.12. In a degree 2 nilpotent group G, establish the formula

$$g^n h^n = (gh)^n [g,h]^{-\binom{n}{2}}$$

for all $g, h \in G$ and $n \in \mathbf{Z}$. This is the first non-trivial case of the *Hall-Petresco formula*, a discrete analogue of the Baker-Campbell-Hausdorff formula that expresses the polynomial sequence $n \mapsto g^n h^n$ explicitly in the form (1.51).

Define a *nilpotent filtered Lie group* of degree $\leq s$ to be a nilpotent filtered group of degree $\leq s$, in which $G = G_{\geq 0}$ and all of the $G_{\geq i}$ are connected, simply connected finite-dimensional Lie groups. A model example here is the *Heisenberg group*, which is the degree 2 nilpotent filtered Lie group

$$G = G_{\geq 0} = G_{\geq 1} := \begin{pmatrix} 1 & \mathbf{R} & \mathbf{R} \\ 0 & 1 & \mathbf{R} \\ 0 & 0 & 1 \end{pmatrix}$$

1.6. Inverse conjecture over the integers

(i.e., the group of upper-triangular unipotent matrices with arbitrary real entries in the upper triangular positions) with

$$G_{\geq 2} := \begin{pmatrix} 1 & 0 & \mathbf{R} \\ 0 & 1 & 0 \\ 0 & 0 & 1 \end{pmatrix}$$

and $G_{\geq i}$ trivial for $i > 2$ (so in this case, $G_{\geq i}$ is also the lower central series).

Exercise 1.6.13. Show that a sequence

$$g(n) = \begin{pmatrix} 1 & x(n) & y(n) \\ 0 & 1 & z(n) \\ 0 & 0 & 1 \end{pmatrix}$$

from $\mathbf{Z}$ to the Heisenberg group G is a polynomial sequence if and only if x, z are linear polynomials and z is a quadratic polynomial.

It is a standard fact in the theory of Lie groups that a connected, simply connected nilpotent Lie group G is topologically equivalent to its Lie algebra $\mathfrak{g}$, with the homeomorphism given by the exponential map $\exp\colon \mathfrak{g} \to G$ (or its inverse, the logarithm function $\log\colon G \to \mathfrak{g}$). Indeed, the Baker-Campbell-Hausdorff formula lets one use the nilpotent Lie algebra $\mathfrak{g}$ to build a connected, simply connected Lie group with that Lie algebra, which is then necessarily isomorphic to G. One can thus classify filtered nilpotent Lie groups in terms of filtered nilpotent Lie algebras, i.e., a nilpotent Lie algebra $\mathfrak{g} = \mathfrak{g}_{\geq 0}$ together with a nested family of sub-Lie algebras

$$\mathfrak{g}_{\geq 0} \geq \mathfrak{g}_{\geq 1} \geq \cdots \geq \mathfrak{g}_{\geq s+1} = \{0\}$$

with the inclusions $[\mathfrak{g}_i, \mathfrak{g}_j] \subset \mathfrak{g}_{i+j}$ (in which the bracket is now the Lie bracket rather than the commutator). One can describe such filtered nilpotent Lie algebras even more precisely using *Mal'cev bases*; see [**Ma1949**], [**Le2005**]. For instance, in the case of the Heisenberg group, one has

$$\mathfrak{g} = \mathfrak{g}_{\geq 0} = \mathfrak{g}_{\geq 1} := \begin{pmatrix} 0 & \mathbf{R} & \mathbf{R} \\ 0 & 0 & \mathbf{R} \\ 0 & 0 & 0 \end{pmatrix}$$

and

$$\mathfrak{g}_{\geq 2} := \begin{pmatrix} 0 & 0 & \mathbf{R} \\ 0 & 0 & 0 \\ 0 & 0 & 0 \end{pmatrix}.$$

From the filtration property, we see that for $i \geq 0$, each $G_{\geq i+1}$ is a normal closed subgroup of $G_{\geq i}$, and for $i \geq 1$, the quotient group $G_{\geq i+1}/G_{\geq i}$ is a connected, simply connected abelian Lie group (with Lie algebra $\mathfrak{g}_{\geq i+1}/\mathfrak{g}_{\geq i}$), and is thus isomorphic to a vector space (with the additive group law). Related to this, one can view $G = G_{\geq 0}$ as a group extension of the quotient group $G/G_{>s}$ (with the degree $s-1$ filtration $(G_{\geq i}/G_{\geq s})$) by the central vector space $G_{\geq s}$. Thus one can view degree s filtered nilpotent groups as an s-fold iterated tower of central extensions by finite-dimensional vector spaces starting from a point; for instance, the Heisenberg group is an extension of $\mathbf{R}^2$ by $\mathbf{R}$.

We thus see that nilpotent filtered Lie groups are generalisations of vector spaces (which correspond to the degree 1 case). We now turn to filtered *nilmanifolds*, which are generalisations of tori. A degree s filtered nilmanifold $G/\Gamma = (G/\Gamma, G_\bullet, \Gamma)$ is a filtered degree s nilpotent Lie group $G_\bullet$, together with a discrete subgroup Γ of G, such that all the subgroups $G_{\geq i}$ in the filtration are *rational* relative to Γ, which means that the subgroup $\Gamma_{\geq i} := \Gamma \cap G_{\geq i}$ is a cocompact subgroup of $G_{\geq i}$ (i.e., the quotient space $G_{\geq i}/\Gamma_{\geq i}$ is cocompact, or equivalently one can write $G_{\geq i} = \Gamma_{\geq i} \cdot K_{\geq i}$ for some compact subset $K_{\geq i}$ of $G_{\geq i}$. Note that the subgroups $\Gamma_{\geq i}$ give Γ the structure of a degree s filtered nilpotent group $\Gamma_\bullet$.

Exercise 1.6.14. Let $G := \mathbf{R}^2$ and $\Gamma := \mathbf{Z}^2$, and let $\alpha \in \mathbf{R}$. Show that the subgroup $\{(x, \alpha x) : x \in \mathbf{R}\}$ of G is rational relative to Γ if and only if α is a rational number; this may help explain the terminology "rational".

By hypothesis, the quotient space $G/\Gamma = G_{\geq 0}/\Gamma_{\geq 0}$ is a smooth compact manifold. The space $G_{\geq s}/\Gamma_{\geq s}$ is a compact connected abelian Lie group, and is thus a torus; the degree s filtered nilmanifold G/Γ can then be viewed as a principal torus bundle over the degree $s-1$ filtered nilmanifold $G/(G_{\geq s}\Gamma)$ with $G_{\geq s}/\Gamma_{\geq s}$ as the structure group; thus one can view degree s filtered nilmanifolds as an s-fold iterated tower of torus extensions starting from a point. For instance, the *Heisenberg nilmanifold*

$$G/\Gamma := \begin{pmatrix} 1 & \mathbf{R} & \mathbf{R} \\ 0 & 1 & \mathbf{R} \\ 0 & 0 & 1 \end{pmatrix} / \begin{pmatrix} 1 & \mathbf{Z} & \mathbf{Z} \\ 0 & 1 & \mathbf{Z} \\ 0 & 0 & 1 \end{pmatrix}$$

is an extension of the two-dimensional torus $\mathbf{R}^2/\mathbf{Z}^2$ by the circle $\mathbf{R}/\mathbf{Z}$.

Every torus of some dimension d can be viewed as a unit cube $[0,1]^d$ with opposite faces glued together; up to measure zero sets, the cube then serves as a fundamental domain for the nilmanifold. Nilmanifolds can be viewed the same way, but the gluing can be somewhat "twisted":

Exercise 1.6.15. Let G/Γ be the Heisenberg nilmanifold. If we abbreviate

$$[x, y, z] := \begin{pmatrix} 1 & x & y \\ 0 & 1 & z \\ 0 & 0 & 1 \end{pmatrix} \Gamma \in G/\Gamma$$

for all $x, y, z \in \mathbf{R}$, show that for almost all x, y, z, that $[x, y, z]$ has exactly one representation of the form $[a, b, c]$ with $a, b, c \in [0, 1]$, which is given by

1.6. Inverse conjecture over the integers

the identity

$$[x, y, z] = [\{x\}, \{y - x\lfloor z \rfloor\}, \{z\}]$$

where $\lfloor x \rfloor$ is the greatest integer part of x, and $\{x\} := x - \lfloor x \rfloor \in [0, 1)$ is the fractional part function. Conclude that G/Γ is topologically equivalent to the unit cube $[0, 1]^3$ quotiented by the identifications

$$(0, y, z) \sim (1, y, z),$$
$$(x, 0, z) \sim (x, 1, z),$$
$$(x, y, 0) \sim (x, \{y - x\}, 1)$$

between opposite faces.

Note that by using the projection $(x, y, z) \mapsto (x, z)$, we can view the Heisenberg nilmanifold G/Γ as a twisted circle bundle over $(\mathbf{R}/\mathbf{Z})^2$, with the fibers being isomorphic to the unit circle $\mathbf{R}/\mathbf{Z}$. Show that G/Γ is *not* homeomorphic to $(\mathbf{R}/\mathbf{Z})^3$. (*Hint:* Show that there are some non-trivial homotopies between loops that force the fundamental group of G/Γ to be smaller than $\mathbf{Z}^3$.)

The logarithm $\log(\Gamma)$ of the discrete cocompact subgroup Γ can be shown to be a lattice of the Lie algebra $\mathfrak{g}$. After a change of basis, one can thus view the latter algebra as a standard vector space $\mathbf{R}^d$ and the lattice as $\mathbf{Z}^d$. Denoting the standard generators of the lattice (and the standard basis of $\mathbf{R}^d$) as $e_1, \ldots, e_d$, we then see that the Lie bracket $[e_i, e_j]$ of two such generators must be an integer combination of more generators:

$$[e_i, e_j] = \sum_{k=1}^{d} c_{ijk} e_k.$$

The *structure constants* c_{ijk} describe completely the Lie group structure of G and Γ. The rational subgroups $G_{\geq l}$ can also be described by picking some generators for $\log(\Gamma_{\geq i})$, which are integer combinations of the $e_1, \ldots, e_d$. We say that the filtered nilmanifold has *complexity* at most M if the dimension and degree is at most M, and the structure constants and coefficients of the generators also have magnitude at most M. This is an admittedly artificial definition, but for quantitative applications it is necessary to have *some* means to quantify the complexity of a nilmanifold.

A *polynomial orbit* in a filtered nilmanifold G/Γ is a map $\mathcal{O} \colon \mathbf{Z} \to G/\Gamma$ of the form $\mathcal{O}(n) := g(n)\Gamma$, where $g \colon \mathbf{Z} \to G$ is a polynomial sequence. For instance, any linear orbit $\mathcal{O}(n) = g^n x$, where $x \in G/\Gamma$ and $g \in G$, is a polynomial orbit.

Exercise 1.6.16. For any $\alpha, \beta \in \mathbf{R}$, show that the sequence

$$n \mapsto [\{-\alpha n\}, \{\alpha n \lfloor \beta n \rfloor\}, \{\beta n\}]$$

(using the notation from Exercise 1.6.15) is a polynomial sequence in the Heisenberg nilmanifold.

With the above example, we see the emergence of *bracket polynomials* when representing polynomial orbits in a fundamental domain. Indeed, one can view the entire machinery of orbits in nilmanifolds as a means of efficiently capturing such polynomials in an algebraically tractable framework (namely, that of polynomial sequences in nilpotent groups). The piecewise continuous nature of the bracket polynomials is then ultimately tied to the twisted gluing needed to identify the fundamental domain with the nilmanifold.

Finally, we can define the notion of a (basic Lipschitz) nilsequence of degree $\leq s$. This is a sequence $\psi\colon \mathbf{Z} \to \mathbf{C}$ of the form $\psi(n) := F(\mathcal{O}(n))$, where $\mathcal{O}\colon \mathbf{Z} \to G/\Gamma$ is a polynomial orbit in a filtered nilmanifold of degree $\leq s$, and $F\colon G/\Gamma \to \mathbf{C}$ is a Lipschitz[21] function. We say that the nilsequence has *complexity* at most M if the filtered nilmanifold has complexity at most M, and the (inhomogeneous Lipschitz norm) of F is also at most M.

A basic example of a degree $\leq s$ nilsequence is a polynomial phase $n \mapsto e(P(n))$, where $P\colon \mathbf{Z} \to \mathbf{R}/\mathbf{Z}$ is a polynomial of degree $\leq s$. A bit more generally, $n \mapsto F(P(n))$ is a degree $\leq s$ sequence, whenever $F\colon \mathbf{R}/\mathbf{Z} \to \mathbf{C}$ is a Lipschitz function. In view of Exercises 1.6.15, 1.6.16, we also see that

(1.52) $$n \mapsto e(\alpha n \lfloor \beta n \rfloor)\psi(\{\alpha n\})\psi(\{\beta n\})$$

or, more generally,

$$n \mapsto F(\alpha n \lfloor \beta n \rfloor)\psi(\{\alpha n\})\psi(\{\beta n\})$$

are also degree ≤ 2 nilsequences, where $\psi\colon [0,1] \to \mathbf{C}$ is a Lipschitz function that vanishes near 0 and 1. The $\psi(\{\alpha n\})$ factor is not needed (as there is no twisting in the x coordinate in Exercise 1.6.15), but the $\psi(\{\beta n\})$ factor is (unfortunately) necessary, as otherwise one encounters the discontinuity inherent in the $\lfloor \beta n \rfloor$ term (and one would merely have a *piecewise Lipschitz* nilsequence rather than a genuinely Lipschitz nilsequence). Because of this discontinuity, bracket polynomial phases $n \mapsto e(\alpha n \lfloor \beta n \rfloor)$ cannot quite be viewed as Lipschitz nilsequences, but from a heuristic viewpoint it is often helpful to pretend as if bracket polynomial phases are model instances of nilsequences.

The only degree ≤ 0 nilsequences are the constants. The degree ≤ 1 nilsequences are essentially the quasiperiodic functions:

---

[21] One needs a metric on G/Γ to define the Lipschitz constant, but this can be done, for instance, by using a basis $e_1, \ldots, e_d$ of Γ to identify G/Γ with a fundamental domain $[0,1]^d$, and using this to construct some (artificial) metric on G/Γ. The details of such a construction will not be important here.

1.6. Inverse conjecture over the integers

Exercise 1.6.17. Show that a degree ≤ 1 nilsequence of complexity M is Fourier-measurable with growth function $\mathcal{F}_M$ depending only on M, where Fourier measurability was defined in Section 1.2.

Exercise 1.6.18. Show that the class of nilsequences of degree $\leq s$ does not change if we drop the condition $G = G_{\leq 0}$, or if we add the additional condition $G = G_{\leq 1}$.

Remark 1.6.11. The space of nilsequences is also unchanged if one insists that the polynomial orbit be linear, and that the filtration be the lower central series filtration; and this is in fact the original definition of a nilsequence. The proof of this equivalence is a little tricky, though; see [**GrTaZi2010b**].

1.6.3. Connection with the Gowers norms.
We define the Gowers norm $\|f\|_{U^d[N]}$ of a function $f \colon [N] \to \mathbf{C}$ by the formula

$$\|f\|_{U^d[N]} := \|f\|_{U^d(\mathbf{Z}/N'\mathbf{Z})} / \|1_{[N]}\|_{U^d(\mathbf{Z}/N'\mathbf{Z})}$$

where N' is any integer greater than $(d+1)N$, $[N]$ is embedded inside $\mathbf{Z}/N'\mathbf{Z}$, and f is extended by zero outside of $[N]$. It is easy to see that this definition is independent of the choice of N'. Note also that the normalisation factor $\|1_{[N]}\|_{U^d(\mathbf{Z}/N'\mathbf{Z})}$ is comparable to 1 when d is fixed and N' is comparable to N.

One of the main reasons why nilsequences are relevant to the theory of the Gowers norms is that they are an obstruction to that norm being small. More precisely, we have

Theorem 1.6.12 (Converse to the inverse conjecture for the Gowers norms). *Let $f \colon [N] \to \mathbf{C}$ be such that $\|f\|_{L^\infty[N]} \leq 1$ and $|\langle f, \psi \rangle_{L^2([N])}| \geq \delta$ for some degree $\leq s$ nilsequence of complexity at most M. Then $\|f\|_{U^{s+1}[N]} \gg_{s,\delta,M} 1$.*

We now prove this theorem, using an argument from [**GrTaZi2009**]. It is convenient to introduce a few more notions. Define a *vertical character* of a degree $\leq s$ filtered nilmanifold G/Γ to be a continuous homomorphism $\eta \colon G_{\geq s} \to \mathbf{R}/\mathbf{Z}$ that annihilates $\Gamma_{\geq s}$, or equivalently an element of the Pontryagin dual $\widehat{G_{\geq s}/\Gamma_{\geq s}}$ of the torus $G_{\geq s}/\Gamma_{\geq s}$. A function $F \colon G/\Gamma \to \mathbf{C}$ is said to have *vertical frequency* η if F obeys the equation

$$F(g_s x) = e(\eta(g_s)) F(x)$$

for all $g_s \in G_{\geq s}$ and $x \in G/\Gamma$. A degree $\leq s$ nilsequence is said to *have a vertical frequency* if it can be represented in the form $n \mapsto F(\mathcal{O}(n))$ for some Lipschitz F with a vertical frequency.

For instance, a polynomial phase $n \mapsto e(P(n))$, where $P \colon \mathbf{Z} \to \mathbf{R}/\mathbf{Z}$ is a polynomial of degree $\leq s$, is a degree $\leq s$ nilsequence with a vertical frequency. Any nilsequence of degree $\leq s - 1$ is trivially a nilsequence of degree $\leq s$ with a vertical frequency of 0. Finally, observe that the space of degree $\leq s$ nilsequences with a vertical frequency is closed under multiplication and complex conjugation.

Exercise 1.6.19. Show that a degree ≤ 1 nilsequence with a vertical frequency necessarily takes the form $\psi(n) = ce(\alpha n)$ for some $c \in \mathbf{C}$ and $\alpha \in \mathbf{R}$ (and conversely, all such sequences are degree ≤ 1 nilsequences with a vertical frequency). Thus, up to constants, degree ≤ 1 nilsequences with a vertical frequency are the same as Fourier characters.

A basic fact (generalising the invertibility of the Fourier transform in the degree ≤ 1 case) is that the nilsequences with vertical frequency generate all the other nilsequences:

Exercise 1.6.20. Show that any degree $\leq s$ nilsequence can be approximated to arbitrary accuracy in the uniform norm by a linear combination of nilsequences with a vertical frequency. (*Hint:* Use the *Stone-Weierstrass theorem.*)

More quantitatively, show that a degree $\leq s$ nilsequence of complexity $\leq M$ can be approximated uniformly to error ε by a sum of $O_{M,\varepsilon,s}(1)$ nilsequences, each with a representation with a vertical frequency that is of complexity $O_{M,\varepsilon,s}(1)$. (*Hint:* This can be deduced from the qualitative result by a compactness argument using the *Arzelá-Ascoli theorem.*)

A derivative $\Delta_h e(P(n))$ of a polynomial phase is a polynomial phase of one lower degree. There is an analogous fact for nilsequences with a vertical frequency:

Lemma 1.6.13 (Differentiating nilsequences with a vertical frequency). *Let $s \geq 1$, and let ψ be a degree $\leq s$ nilsequence with a vertical frequency. Then for any $h \in \mathbf{Z}$, $\Delta_h \psi$ is a degree $\leq s-1$ nilsequence. Furthermore, if ψ has complexity $\leq M$ (with a vertical frequency representation), then $\Delta_h \psi$ has complexity $O_{M,s}(1)$.*

Proof. We just prove the first claim, as the second claim follows by refining the argument.

We write $\psi = F(g(n)\Gamma)$ for some polynomial sequence $g\colon \mathbf{Z} \to G/\Gamma$ and some Lipschitz function F with a vertical frequency. We then express

$$\Delta_h \psi(n) = \tilde{F}(\tilde{g}(n)(\Gamma \times \Gamma))$$

where $\tilde{F}\colon G \times G/(\Gamma \times \Gamma) \to \mathbf{C}$ is the function

$$\tilde{F}(x,y) := F(x)\overline{F(y)}$$

1.6. Inverse conjecture over the integers

and $\tilde{g} \colon \mathbf{Z} \to G \times G$ is the sequence

$$\tilde{g}(n) := (g(n), \partial_h g(n) g(n)).$$

Now we give a filtration on $G \times G$ by setting

$$(G \times G)_{\geq j} := G_{\geq j} \times_{G_{\geq j+1}} G_{\geq j}$$

for $j \geq 0$, where $G_{\geq j} \times_{G_{\geq j+1}} G_{\geq j}$ is the subgroup of $G_{\geq j} \times_{G_{\geq j+1}} G_{\geq j}$ generated by $G_{\geq j+1} \times G_{\geq j+1}$ and the diagonal group $G^\Delta_{\geq j} := \{(g_j, g_j) : g_j \in G_{\geq j}\}$. One easily verifies that this is a filtration on $G \times G$. The sequences $(g(n), g(n))$ and $(\mathrm{id}, \partial_h g(n))$ are both polynomial with respect to this filtration, and hence by the Lazard-Leibman theorem (Theorem 1.6.8), $\tilde{g}$ is polynomial also.

Next, we use the hypothesis that F has a vertical frequency to conclude that $\tilde{F}$ is invariant with respect to the action of the diagonal group $G^\Delta_s = (G \times G)_{\geq s}$. If we then define $G^\square$ to be the Lie group $G^\square := (G \times G)_{\geq 0}/G^\Delta_s$ with filtration $G^\square_{\geq j} := (G \times G)_{\geq j}/G^\Delta_s$, then $G^\square$ is a degree $\leq s-1$ filtered nilpotent Lie group; setting $\Gamma^\square := (\Gamma \times \Gamma) \cap G^\square$, we conclude that $G^\square/\Gamma^\square$ is a degree $\leq s-1$ nilmanifold and

$$\Delta_h \psi(n) = F^\square(g^\square(n)\Gamma^\square)$$

where $F^\square, g^\square$ are the projections of $\tilde{F}, \tilde{g}$ from $G \times G$ to $G^\square$. The claim follows. $\square$

We now prove Theorem 1.6.12 by induction on s. The claim is trivial for $s = 0$, so we assume that $s \geq 1$ and that the claim has already been proven for smaller values of s.

Let f, δ, ψ be as in Theorem 1.6.12. From Exercise 1.6.20 we see (after modifying δ, M) that we may assume that ψ has a vertical frequency. Next, we use the identity

$$|\mathbf{E}_{n \in \mathbf{Z}/N\mathbf{Z}'} f(n)\overline{\psi(n)}|^2 = \mathbf{E}_{h \in \mathbf{Z}/N'\mathbf{Z}} \mathbf{E}_{n \in \mathbf{Z}/N'\mathbf{Z}} \Delta_h f(n) \overline{\Delta_h \psi(n)}$$

(extending f by zero outside of $[N]$, and extending ψ arbitrarily) to conclude that

$$|\mathbf{E}_{n \in [N]} \Delta_h f(n) \overline{\Delta_h \psi(n)}| \gg_\delta 1$$

for $\gg N$ values of $h \in [-N, N]$. By induction hypothesis and Lemma 1.6.13, we conclude that

$$\|\Delta_h f\|_{U^s[N]} \gg_{\delta, M} 1$$

for $\gg N$ values of $h \in [-N, N]$. Using the identity

$$\|f\|^{2^{s+1}}_{U^{s+1}(\mathbf{Z}/N'\mathbf{Z})} = \mathbf{E}_{h \in \mathbf{Z}/N'\mathbf{Z}} \|\Delta_h f\|^{2^s}_{U^s(\mathbf{Z}/N'\mathbf{Z})}$$

we close the induction and obtain the claim.

In the other direction, we have the following recent result:

Theorem 1.6.14 (Inverse conjecture for the Gowers norms on $\mathbf{Z}$). *(See* [**GrTaZi2010b**]*.) Let $f\colon [N] \to \mathbf{C}$ be such that $\|f\|_{L^\infty[N]} \leq 1$ and $\|f\|_{U^{s+1}[N]} \geq \delta$. Then $|\langle f, \psi\rangle_{L^2([N])}| \gg_{s,\delta} 1$ for some degree $\leq s$ nilsequence of complexity $O_{s,\delta}(1)$.*

An extensive heuristic discussion of how this conjecture is proven can be found in [**GrTaZi2010**]; for the purposes of this text, we shall simply accept this theorem as a black box. For a discussion of the history of the conjecture, including the cases $s \leq 3$; see [**GrTaZi2009**]. An alternate proof to Theorem 1.6.14 was recently also established in [**CaSz2010**], [**Sz2010b**]. These methods are based on the proof of an analogous ergodic-theory result to Theorem 1.6.14, namely the description of the characteristic factors for the Gowers-Host-Kra semi-norms in [**HoKr2005**], which we will not discuss here, except to say that one of the main ideas is to construct, and then study, spaces analogous to the Host-Kra groups $\mathrm{HK}^k(G)$ and Host-Kra nilmanifolds $\mathrm{HK}^k(G)/\mathrm{HK}^k(\Gamma)$ associated to an arbitrary function f on a (limit-finite) interval $[N]$ (or of a function f on an ergodic measure-preserving system).

Exercise 1.6.21 (99% inverse theorem).

(i) (Straightening an approximately linear function) Let $\varepsilon, \kappa > 0$. Let $\xi\colon [-N,N] \to \mathbf{R}/\mathbf{Z}$ be a function such that $|\xi(a+b) - \xi(a) - \xi(b)| \leq \kappa$ for all but εN^2 of all $a, b \in [-N, N]$ with $a+b \in [-N, N]$. If ε is sufficiently small, show that there exists an affine linear function $n \mapsto \alpha n + \beta$ with $\alpha, \beta \in \mathbf{R}/\mathbf{Z}$ such that $|\xi(n) - \alpha n - \beta| \ll_\varepsilon \kappa$ for all but $\delta(\varepsilon)N$ values of $n \in [-N, N]$, where $\delta(\varepsilon) \to 0$ as $\varepsilon \to 0$. (*Hint:* One can take κ to be small. First find a way to lift ξ in a nice manner from $\mathbf{R}/\mathbf{Z}$ to $\mathbf{R}$.)

(ii) Let $f\colon [N] \to \mathbf{C}$ be such that $\|f\|_{L^\infty[N]} \leq 1$ and $\|f\|_{U^{s+1}[N]} \geq 1-\varepsilon$. Show that there exists a polynomial $P\colon \mathbf{Z} \to \mathbf{R}/\mathbf{Z}$ of degree $\leq s$ such that $\|f - e(P)\|_{L^2([N])} \leq \delta$, where $\delta = \delta_s(\varepsilon) \to 0$ as $\varepsilon \to 0$ (holding s fixed). *Hint:* Adapt the argument of the analogous finite field statement. One cannot exploit the discrete nature of polynomials any longer; and so one must use the preceding part of the exercise as a substitute.

The inverse conjecture for the Gowers norms, when combined with the equidistribution theory for nilsequences that we will turn to next, has a number of consequences, analogous to the consequences for the finite field analogues of these facts; see [**GrTa2010b**] for further discussion.

1.6.4. Equidistribution of nilsequences. In the subject of higher order Fourier analysis and, in particular, in the proof of the inverse conjecture for the Gowers norms, as well as in several of the applications of this conjecture,

1.6. Inverse conjecture over the integers

it will be of importance to be able to compute statistics of nilsequences ψ, such as their averages $\mathbf{E}_{n \in [N]} \psi(n)$ for a large integer N; this generalises the computation of exponential sums such as $\mathbf{E}_{n \in [N]} e(P(n))$ that occurred in Section 1.1. This is closely related to the equidistribution of polynomial orbits $\mathcal{O} \colon \mathbf{Z} \to G/\Gamma$ in nilmanifolds. Note that as G/Γ is a compact quotient of a locally compact group G, it comes endowed with a unique left-invariant Haar measure $\mu_{G/\Gamma}$ (which is isomorphic to the Lebesgue measure on a fundamental domain $[0,1]^d$ of that nilmanifold). By default, when we talk about equidistribution in a nilmanifold, we mean with respect to the Haar measure; thus $\mathcal{O}$ is asymptotically equidistributed if and only if

$$\lim_{N \to \infty} \mathbf{E}_{n \in [N]} F(\mathcal{O}(n)) = 0$$

for all Lipschitz $F \colon G/\Gamma \to \mathbf{C}$. One can also describe single-scale equidistribution (and non-standard equidistribution) in a similar fashion, but for the sake of discussion let us restrict our attention to the simpler and more classical situation of asymptotic equidistribution here (although it is the single-scale equidistribution theory which is ultimately relevant to questions relating to the Gowers norms).

When studying equidistribution of polynomial sequences in a torus $\mathbf{T}^d$, a key tool was the *van der Corput lemma* (Lemma 1.1.6). This lemma asserted that if a sequence $x \colon \mathbf{Z} \to \mathbf{T}^d$ is such that all derivatives $\partial_h x \colon \mathbf{Z} \to \mathbf{T}^d$ with $h \neq 0$ are asymptotically equidistributed, then x itself is also asymptotically equidistributed.

The notion of a derivative requires the ability to perform subtraction on the range space $\mathbf{T}^d$: $\partial_h x(n+h) - \partial_h x(n)$. When working in a higher degree nilmanifold G/Γ, which is not a torus, we do not have a notion of subtraction. However, such manifolds are still torus *bundles* with torus $\mathbf{T} := G_{\geq s}/\Gamma_{\geq s}$. This gives a weaker notion of subtraction, namely the map $\pi \colon G/\Gamma \times G/\Gamma \to (G/\Gamma \times G/\Gamma)/\mathbf{T}^\Delta$, where $\mathbf{T}^\Delta$ is the diagonal action $g_s \colon (x, y) \mapsto (g_s x, g_s y)$ of the torus $\mathbf{T}$ on the product space $G/\Gamma \times G/\Gamma$. This leads to a generalisation of the van der Corput lemma:

Lemma 1.6.15 (Relative van der Corput lemma). *Let $x \colon \mathbf{Z} \to G/\Gamma$ be a sequence in a degree $\leq s$ nilmanifold for some $s \geq 1$. Suppose that the projection of x to the degree $\leq s-1$ filtered nilmanifold $G/G_s\Gamma$ is asymptotically equidistributed, and suppose also that for each non-zero $h \in \mathbf{Z}$, the sequence $\partial_h x \colon n \mapsto \pi(x(n+h), x(n))$ is asymptotically equidistributed with respect to some $\mathbf{T}$-invariant measure μ_h on $(G/\Gamma \times G/\Gamma)/\mathbf{T}^\Delta$. Then x is asymptotically equidistributed in G/Γ.*

Proof. It suffices to show that, for each Lipschitz function $F\colon G/\Gamma \to \mathbf{C}$, that
$$\lim_{n\to\infty} \mathbf{E}_{n\in[N]} F(x(n)) = \int_{G/\Gamma} F\, d\mu_{G/\Gamma}.$$

By Exercise 1.6.20, we may assume that F has a vertical frequency. If this vertical frequency is non-zero, then F descends to a function on the degree $\leq s-1$ filtered nilmanifold $G/G_s\Gamma$, and the claim then follows from the equidistribution hypothesis on this space. So suppose instead that F has a non-zero vertical frequency. By vertically rotating F (and using the G_s-invariance of $\mu_{G/\Gamma}$ we conclude that $\int_{G/\Gamma} F\mu_{G/\Gamma} = 0$. Applying the van der Corput inequality (Lemma 1.1.6), we now see that it suffices to show that
$$\lim_{n\to\infty} \mathbf{E}_{n\in[N]} F(x(n+h))\overline{F(x(n))} = 0$$
for each non-zero h. The function $(x,y) \to F(x)\overline{F(y)}$ on $G/\Gamma \times G/\Gamma$ is T^Δ-invariant (because of the vertical frequency hypothesis) and so descends to a function $\tilde F$ on $(G/\Gamma \times G/\Gamma)/T^\Delta$. We thus have
$$\lim_{n\to\infty} \mathbf{E}_{n\in[N]} F(x(n+h))\overline{F(x(n))} = \int_{(G/\Gamma \times G/\Gamma)/T^\Delta} \tilde F\, d\mu_h.$$

The function $\tilde F$ has a non-zero vertical frequency with respect to the residual action of $\mathbf{T}$ (or more precisely, of $(\mathbf{T}\times\mathbf{T})/\mathbf{T}^\Delta$, which is isomorphic to $\mathbf{T}$). As μ_h is invariant with respect to this action, the integral thus vanishes, as required. $\square$

This gives a useful criterion for equidistribution of polynomial orbits. Define a *horizontal character* to be a continuous homomorphism η from G to $\mathbf{R}/\mathbf{Z}$ that annihilates Γ (or equivalently, an element of the Pontryagin dual of the *horizontal torus* $G/([G,G]\Gamma)$. This is easily seen to be a torus. Let $\pi_i\colon G_{\geq i} \to \mathbf{T}_i$ be the projection map.

Theorem 1.6.16 (Leibman equidistribution criterion). *Let $\mathcal{O}\colon n \mapsto g(n)\Gamma$ be a polynomial orbit on a degree $\leq s$ filtered nilmanifold G/Γ. Suppose that $G = G_{\geq 0} = G_{\geq 1}$. Then $\mathcal{O}$ is asymptotically equidistributed in G/Γ if and only if $\eta \circ g$ is non-constant for each non-trivial horizontal character.*

This theorem was first established by Leibman [**Le2005**] (by a slightly different method), and also follows from the above van der Corput lemma and some tedious additional computations; see [**GrTa2011**] for details. For linear orbits, this result was established in [**Pa1970**], [**Gr1961**]. Using this criterion (together with more quantitative analogues for single-scale equidistribution), one can develop equidistribution decompositions that generalise those in Section 1.1. Again, the details are technical and we will refer to

[GrTa2011] for details. We give a special case of Theorem 1.6.16 as an exercise:

Exercise 1.6.22. Use Lemma 1.6.15 to show that if α, β are two real numbers that are linearly independent modulo 1 over the integers, then the polynomial orbit

$$n \mapsto \begin{pmatrix} 1 & \alpha n & 0 \\ 0 & 1 & \beta n \\ 0 & 0 & 1 \end{pmatrix} \Gamma$$

is asymptotically equidistributed in the Heisenberg nilmanifold G/Γ; note that this is a special case of Theorem 1.6.16. Conclude that the map $n \mapsto \alpha n \lfloor \beta n \rfloor \mod 1$ is asymptotically equidistributed in the unit circle.

One application of this equidistribution theory is to show that bracket polynomial objects such as (1.52) have a negligible correlation with any genuinely quadratic phase $n \mapsto e(\alpha n^2 + \beta n + \gamma)$ (or more generally, with any genuinely polynomial phase of bounded degree); this result was first established in [**Ha1993**]. On the other hand, from Theorem 1.6.12 we know that (1.52) has a large $U^3[N]$ norm. This shows that even when $s = 2$, one cannot invert the Gowers norm purely using polynomial phases. This observation first appeared in [**Go1998**] (with a related observation in [**FuWi1996**]).

Exercise 1.6.23. Let the notation be as in Exercise 1.6.22. Show that

$$\lim_{n \to \infty} \mathbf{E}_{n \in [N]} e(\alpha n \lfloor \beta n \rfloor - \gamma n^2 - \delta n) = 0$$

for any $\gamma, \delta \in \mathbf{R}$. (*Hint:* You can either apply Theorem 1.6.16, or go back to Lemma 1.6.15.)

1.7. Linear equations in primes

In this section, we discuss one of the motivating applications of the theory developed thus far, namely to count solutions to linear equations in primes $\mathcal{P} = \{2, 3, 5, 7, \ldots\}$ (or in dense subsets A of primes $\mathcal{P}$). Unfortunately, the most famous linear equations in primes, the twin prime equation $p_2 - p_1 = 2$ and the even Goldbach equation $p_1 + p_2 = N$, remain out of reach of this technology (because the relevant affine linear forms involved are commensurate, and thus have infinite complexity with respect to the Gowers norms), but most other systems of equations, in particular, that of arithmetic progressions $p_i = n + ir$ for $i = 0, \ldots, k - 1$ (or equivalently, $p_i + p_{i+2} = 2p_{i+1}$ for $i = 0, \ldots, k-2$), as well as the odd Goldbach equation $p_1 + p_2 + p_3 = N$, are tractable.

To illustrate the main ideas, we will focus on the following result of Green [**Gr2005**]:

Theorem 1.7.1 (Roth's theorem in the primes [**Gr2005**]). *Let $A \subset \mathcal{P}$ be a subset of primes whose upper density $\limsup_{N\to\infty} |A \cap [N]|/|\mathcal{P} \cap [N]|$ is positive. Then A contains infinitely many arithmetic progressions of length three.*

This should be compared with Roth's theorem in the integers (Section 1.2), which is the same statement but with the primes $\mathcal{P}$ replaced by the integers $\mathbf{Z}$ (or natural numbers $\mathbf{N}$). Indeed, Roth's theorem for the primes is proven by *transferring* Roth's theorem for the integers to the prime setting; the latter theorem is used as a "black box". The key difficulty here in performing this transference is that the primes have zero density inside the integers; indeed, from the prime number theorem we have $|\mathcal{P} \cap [N]| = (1 + o(1))\frac{N}{\log N} = o(N)$.

There are a number of generalisations of this transference technique. In [**GrTa2008b**], the above theorem was extended to progressions of longer length (thus transferring Szemerédi's theorem to the primes). In a series of papers [**GrTa2010, GrTa2011, GrTa2008c, GrTaZi2010b**], related methods are also used to obtain an asymptotic for the number of solutions in the primes to any system of linear equations of bounded complexity. This latter result uses the full power of higher order Fourier analysis, in particular, relying heavily on the inverse conjecture for the Gowers norms; in contrast, Roth's theorem and Szemerédi's theorem in the primes are "softer" results that do not need this conjecture.

To transfer results from the integers to the primes, there are three basic steps:

(i) A general transference principle, that transfers certain types of additive combinatorial results from dense subsets of the integers to dense subsets of a suitably "pseudorandom set" of integers (or more precisely, to the integers weighted by a suitably "pseudorandom measure").

(ii) An application of sieve theory to show that the primes (or more precisely, an affine modification of the primes) lie inside a suitably pseudorandom set of integers (or more precisely, have significant mass with respect to a suitably pseudorandom measure).

(iii) If one is seeking asymptotics for patterns in the primes, and not simply lower bounds, one also needs to control correlations between the primes (or proxies for the primes, such as the Möbius function) with various objects that arise from higher order Fourier analysis, such as nilsequences.

1.7. Linear equations in primes

The former step can be accomplished[22] in a number of ways. For progressions of length three (and more generally, for controlling linear patterns of complexity at most one), transference can be accomplished by Fourier-analytic methods. For more complicated patterns, one can use techniques inspired by ergodic theory; more recently, simplified and more efficient methods based on duality (the Hahn-Banach theorem) have also been used. No number theory is used in this step.

The second step is accomplished by fairly standard sieve theory methods (e.g. the Selberg sieve, or the slight variants of this sieve used by Goldston, Pintz, and Yıldırım [**GoPiYi2008**]). Remarkably, very little of the formidable apparatus of modern analytic number theory is needed for this step; for instance, the only fact about the Riemann zeta function that is truly needed is that it has a simple pole at $s=1$, and no knowledge of L-functions is needed.

The third step does draw more significantly on analytic number theory techniques and results (most notably, the method of Vinogradov to compute oscillatory sums over the primes, and also the Siegel-Walfisz theorem that gives a good error term on the prime number theorem in arithemtic progressions). As these techniques are somewhat orthogonal to the main topic of this text, we shall only touch briefly on this aspect of the transference strategy.

1.7.1. Transference. The transference principle is not a single theorem, but is instead a family of related results with a common purpose, namely to show that a sufficiently pseudorandom set, measure, or probability distribution will be "indistinguishable" from the whole set (or the uniform measure or probability distribution) in certain statistical senses. A key tool in this regard is a *dense model theorem* that allows one to *approximate* or *model* any set or function that is dense with respect to a pseudorandom measure, by a set or function which is dense with respect to the uniform measure. It turns out that one can do this as long as the approximation is made with respect to a sufficiently *weak* topology; for the applications to counting arithmetic patterns, it turns out that the topology given by the Gowers norms is the right one to use. The somewhat complicated nature of these norms, though, does make the verification of the required pseudorandomness properties to be slightly tricky.

We illustrate these themes with Roth's theorem, though the general strategy applies to several other results in additive combinatorics. We begin with Roth's theorem in a cyclic group $\mathbf{Z}/N\mathbf{Z}$, which we phrase as follows:

---

[22]In the case of transference to *genuinely* random sets, rather than pseudorandom sets, similar ideas appeared earlier in the graph theory setting; see [**KoLuRo1996**].

Theorem 1.7.2 (Roth's theorem in $\mathbf{Z}/N\mathbf{Z}$). *Let N be odd. If $f\colon \mathbf{Z}/N\mathbf{Z} \to \mathbf{R}$ is a function obeying the pointwise bound $0 \leq f \leq 1$ and the lower bound $\mathbf{E}_{n\in \mathbf{Z}/N\mathbf{Z}}f(n) \geq \delta > 0$, then one has $\Lambda(f,f,f) \geq c(\delta)$ for some $c(\delta) > 0$, where $\Lambda(f,g,h) := \mathbf{E}_{n,r\in \mathbf{Z}/N\mathbf{Z}}f(n)g(n+r)h(n+2r)$.*

We assume this theorem as a "black box", in that we will not care as to how this theorem is proven. As noted in previous sections, this theorem easily implies the existence of non-trivial arithmetic progressions of length three in any subset A of $[N/3]$ (say) with $|A| \geq \delta N$, as long as N is sufficiently large depending on δ, as it provides a non-trivial lower bound on $\Lambda(1_A, 1_A, 1_A)$.

Now we generalise the above theorem. We view N as an (odd) parameter going off to infinity, and use $o_{N\to\infty}(1)$ to denote any quantity that goes to zero as $N \to \infty$. We define a *measure* (or more precisely, a *weight function*) to be a non-negative function $\nu\colon \mathbf{Z}/N\mathbf{Z} \to \mathbf{R}^+$ depending on N, such that $\mathbf{E}_{n\in [N]}\nu(n) = 1 + o_{N\to\infty}(1)$, thus ν is basically the density function of a probability distribution on $\mathbf{Z}/N\mathbf{Z}$. We say that ν is *Roth-pseudorandom* if for every $\delta > 0$ (independent of N) there exists $c_\nu(\delta) > 0$ such that one has the lower bound

$$\Lambda(f,f,f) \geq c_\nu(\delta) + o_{N\to\infty;\delta}(1)$$

whenever $f\colon \mathbf{Z}/N\mathbf{Z} \to \mathbf{R}$ is a function obeying the pointwise bound $0 \leq f \leq \nu$ and the lower bound $\mathbf{E}_{n\in \mathbf{Z}/N\mathbf{Z}}f \geq \delta$, and $o_{N\to\infty;\delta}(1)$ goes to zero as $N \to \infty$ for any fixed δ. Thus, Roth's theorem asserts that the uniform measure 1 is Roth-pseudorandom. Observe that if ν is Roth-pseudorandom, then any subset A of $[N/3]$ whose *weighted density* $\nu(A) := \mathbf{E}_{n\in \mathbf{Z}/N\mathbf{Z}}1_A(n)\nu(n)$ is at least δ will contain a non-trivial arithmetic progression of length three, if N is sufficiently large depending on δ, as we once again obtain a non-trivial lower bound on $\Lambda(1_A, 1_A, 1_A)$ in this case. Thus it is of interest to establish Roth-pseudorandomness for a wide class of measures.

Exercise 1.7.1. Show that if ν is Roth-pseudorandom, and η is another measure which is "uniformly absolutely continuous" with respect to ν in the sense that one has the bound $\eta(A) \leq f(\nu(A)) + o_{N\to\infty}(1)$ all $A \subset \mathbf{Z}/N\mathbf{Z}$ and some function $f\colon \mathbf{R}^+ \to \mathbf{R}^+$ with $f(x) \to 0$ as $x \to 0$, then η is also Roth-pseudorandom.

In view of the above exercise, the case of measures that are absolutely continuous with respect to the uniform distribution is uninteresting: the important case is instead when η is "singular" with respect to the uniform measure, in the sense that it is concentrated on a set of density $o_{N\to\infty}(1)$ with respect to uniform measure, as this will allow us to detect progressions of length three in sparse sets.

1.7. Linear equations in primes

A model example to keep in mind of a candidate for a Roth-pseudorandom measure is a random sparse measure of some small density $0 < p \ll 1$, in which each $\nu(n)$ is an independent random variable that equals $1/p$ with probability p and 0 otherwise. The case $p = 1/\log N$ can be thought of as a crude model for the primes (cf. Cramér's random model for the primes).

Recall that the form $\Lambda(f, g, h)$ is controlled by the U^2 norm in the sense that one has the inequality

$$|\Lambda(f,g,h)| \leq \|f\|_{U^2(\mathbf{Z}/N\mathbf{Z})}$$

whenever $f, g, h \colon \mathbf{Z}/N\mathbf{Z} \to \mathbf{C}$ are bounded in magnitude by 1, and similarly for permutations. Actually one has the slightly more precise inequality

$$|\Lambda(f,g,h)| \leq \|f\|_{u^2(\mathbf{Z}/N\mathbf{Z})}$$

where

$$\|f\|_{u^2(\mathbf{Z}/N\mathbf{Z})} := \sup_{\xi \in \mathbf{Z}/N\mathbf{Z}} |\hat{f}(\xi)|$$

as can easily be seen from the identity

(1.53) $$\Lambda(f,g,h) = \sum_{\xi \in \mathbf{Z}/N\mathbf{Z}} \hat{f}(\xi)\hat{g}(-2\xi)\hat{h}(\xi),$$

Hölder's inequality, and the Plancherel identity.

This suggests a strategy to establish the Roth-pseudorandomness of a measure by showing that functions f dominated by that measure can be approximated in u^2 norm by functions dominated instead by the uniform measure 1. Indeed, we have

Lemma 1.7.3 (Criterion for Roth-pseudorandomness). *Suppose we have a measure ν with the following properties:*

(i) *(Control by u^2) For any $f,g,h \colon \mathbf{Z}/N\mathbf{Z} \to \mathbf{R}$ with the pointwise bound $|f|, |g|, |h| \leq \nu + 1$, one has $|\Lambda(f,g,h)| \leq \alpha(\|f\|_{u^2(\mathbf{Z}/N\mathbf{Z})}) + o_{N \to \infty}(1)$, where $\alpha \colon \mathbf{R}^+ \to \mathbf{R}^+$ is a function with $\alpha(x) \to 0$ as $x \to 0$, and similarly for permutations.*

(ii) *(Approximation in u^2) For any $f \colon \mathbf{Z}/N\mathbf{Z} \to \mathbf{R}$ with the pointwise bound $0 \leq f \leq \nu$, and any $\varepsilon > 0$, there exists $g \colon \mathbf{Z}/N\mathbf{Z} \to \mathbf{R}$ with the pointwise bound $0 \leq g \leq 1 + o_{n \to \infty;\varepsilon}(1)$ such that $\|f - g\|_{u^2(\mathbf{Z}/N\mathbf{Z})} \leq \varepsilon + o_{n \to \infty;\varepsilon}(1)$.*

Then ν is Roth-pseudorandom.

Proof. Let $f \colon \mathbf{Z}/N\mathbf{Z} \to \mathbf{C}$ be such that $0 \leq f \leq \nu$ and $\mathbf{E}_{n \in \mathbf{Z}/N\mathbf{Z}} f \geq \delta$. Let $\varepsilon > 0$ be a small number to be chosen later. We then use the decomposition to split $f = g + (f - g)$ with the above stated properties. Since

$$|\mathbf{E}_{n \in \mathbf{Z}/N\mathbf{Z}} f(n) - g(n)| \leq \|f - g\|_{u^2(\mathbf{Z}/N\mathbf{Z})} \leq \varepsilon + o_{n \to \infty;\varepsilon}(1)$$

we have from the triangle inequality that
$$\mathbf{E}_{n\in\mathbf{Z}/N\mathbf{Z}}g(n) \geq \delta - \varepsilon - o_{n\to\infty;\varepsilon}(1)$$
and, in particular,
$$\mathbf{E}_{n\in\mathbf{Z}/N\mathbf{Z}}g(n) \geq \delta/2$$
for N large enough. Similarly we have $0 \leq g \leq 2$ (say) for N large enough. From Roth's theorem we conclude that
$$\Lambda(g,g,g) \gg c(\delta/4)$$
for N large enough. On the other hand, by the first hypothesis, the other seven terms in
$$\Lambda(f,f,f) = \Lambda(g+(f-g), g+(f-g), g+(f-g))$$
are $O(\alpha(O(\varepsilon)))$ for N large enough. If ε is sufficiently small depending on δ, we obtain the claim. $\square$

Note that this argument in fact gives a value of $c_\nu(\delta)$ that is essentially the same as $c(\delta)$. Also, we see that the u^2 norm here could be replaced by the U^2 norm, or indeed by any other quantity which is strong enough for the control hypothesis to hold, and also weak enough for the approximation property to hold.

So now we need to find some conditions on ν that will allow us to obtain both the control and approximation properties. We begin with the control property. One way to accomplish this is via a restriction estimate:

Lemma 1.7.4 (Restriction estimate implies control). *Let ν be a measure. Suppose there exists an exponent $2 < q < 3$ such that one has the restriction estimate*

(1.54) $$\|\hat{f}\|_{\ell^q(\mathbf{Z}/N\mathbf{Z})} \leq C$$

whenever $f\colon \mathbf{Z}/N\mathbf{Z} \to \mathbf{C}$ obeys the pointwise bound $|f| \leq \nu$, where C is independent of n. Then ν enjoys the control in the u^2 property from Lemma 1.7.3.

Proof. From Plancherel's theorem, we see that (1.54) already holds if we have $|f| \leq 1$, so by the triangle inequality it also holds (with a slightly different value of C) if $|f| \leq \nu + 1$.

Now suppose that $|f|, |g|, |h| \leq \nu+1$. From (1.53) and Hölder's inequality one has
$$|\Lambda(f,g,h)| \leq \|f\|_{\ell^q(\mathbf{Z}/N\mathbf{Z})}^{q-2} \|\hat{f}\|_{\ell^\infty(\mathbf{Z}/N\mathbf{Z})}^{3-q} \|g\|_{\ell^q(\mathbf{Z}/N\mathbf{Z})} \|h\|_{\ell^q(\mathbf{Z}/N\mathbf{Z})}$$
and thus by (1.54)
$$|\Lambda(f,g,h)| \leq C^q \|f\|_{u^2(\mathbf{Z}/N\mathbf{Z})}^{3-q}$$

1.7. Linear equations in primes

and the claim follows. □

Exercise 1.7.2. Show that the estimate (1.54) for $q \leq 2$ can only hold when ν is bounded uniformly in N; this explains the presence of the hypothesis $q > 2$ in the above condition.

Exercise 1.7.3. Show that the estimate (1.54) is equivalent to the estimate

$$\mathbf{E}_{n \in \mathbf{Z}/N\mathbf{Z}} | \sum_{\xi \in \mathbf{Z}/N\mathbf{Z}} g(\xi) e(\xi nx/N) | \nu(n) \leq C \|g\|_{\ell^{q'}(\mathbf{Z}/N\mathbf{Z})}$$

for all $g \colon \mathbf{Z}/N\mathbf{Z} \to \mathbf{C}$, where $q' := q/(q-1)$ is the dual exponent to q. Informally, this asserts that a Fourier series with $\ell^{q'}$ coefficients can be "restricted" to the support of ν in a uniformly absolutely integrable manner (relative to ν). Historically, this is the origin of the term "restriction theorem" (in the context where $\mathbf{Z}/N\mathbf{Z}$ is replaced with a Euclidean space such as $\mathbf{R}^n$, and ν is a surface measure on a manifold such as the sphere S^{n-1}). See for instance [Ta2003].

Now we turn to the approximation property. The approximation g to f needs to be close in u^2 norm, i.e., the Fourier coefficients need to be uniformly close. One attempt to accomplish this is *hard thresholding*: one simply discards all Fourier coefficients in the Fourier expansion

$$f(n) = \sum_{\xi \in \mathbf{Z}/N\mathbf{Z}} \hat{f}(\xi) e(x\xi/N)$$

of f that are too small, thus setting g equal to something like

$$g(n) = \sum_{\xi \in \mathbf{Z}/N\mathbf{Z} : |\hat{f}(\xi)| \geq \varepsilon} \hat{f}(\xi) e(x\xi/N).$$

The main problem with this choice is that there is no guarantee that the non-negativity of f will transfer over to the non-negativity of g; also, there is no particular reason why g would be bounded.

But a small modification of this idea does work, as follows. Let $S := \{\xi \in \mathbf{Z}/N\mathbf{Z} : |\hat{f}(\xi)| \geq \varepsilon\}$ denote the large Fourier coefficients of f. The function g proposed above can be viewed as a convolution $f * K$, where $K(n) := \sum_{\xi \in S} e(x\xi/N)$ and $f * K(n) := \mathbf{E}_{m \in \mathbf{Z}/N\mathbf{Z}} f(m) K(n - m)$. The inability to get good pointwise bounds on $f * K$ can be traced back to the oscillatory nature of the convolution kernel K (which can be viewed as a generalised Dirichlet kernel).

But experience with Fourier analysis tells us that the behaviour of such convolutions improves if one replaces the Dirichlet-type kernels with something more like a Fejér-type kernel instead. With that in mind, we try

$$g(n) := \mathbf{E}_{m_1, m_2 \in B} f(n + m_1 - m_2)$$

where B is the *Bohr set*

$$B := \{n \in \mathbf{Z}/N\mathbf{Z} : |e(n\xi/N) - 1| \leq \varepsilon \text{ for all } \xi \in S\}.$$

Clearly, if f is non-negative, then g is also. Now we look at upper bounds on g. Clearly,

$$g(n) \leq \mathbf{E}_{m_1, m_2 \in B} \nu(n + m_1 - m_2)$$

so by Fourier expansion

$$\|g\|_{L^\infty(\mathbf{Z}/N\mathbf{Z})} \leq \sum_{\xi \in \mathbf{Z}/N\mathbf{Z}} |\mathbf{E}_{m \in B} e(\xi B)|^2 |\hat{\nu}(\xi)|.$$

Let us make the Fourier-pseudorandomness assumption

(1.55)
$$\sup_{\xi \neq 0} |\hat{\nu}(\xi)| = o_{N \to \infty}(1).$$

Evaluating the $\xi = 0$ term on the RHS separately, we conclude that

$$\|g\|_{L^\infty(\mathbf{Z}/N\mathbf{Z})} \leq 1 + o_{N \to \infty}(\sum_{\xi \in \mathbf{Z}/N\mathbf{Z}} |\mathbf{E}_{m \in B} e(\xi B)|^2).$$

By Plancherel's theorem we have

$$\sum_{\xi \in \mathbf{Z}/N\mathbf{Z}} |\mathbf{E}_{m \in B} e(\xi B)|^2 = |B|/N.$$

From the Kronecker approximation theorem we have

$$|B|/N \gg (\varepsilon/10)^{|S|}$$

(say). Finally, if we assume (1.54) we have $|S| \ll \varepsilon^{-q}$. Putting this all together we obtain the pointwise bound

$$g \leq 1 + o_{N \to \infty; q, \varepsilon}(1).$$

Finally, we see how g approximates f. From Fourier analysis one has

$$\hat{g}(\xi) = \hat{f}(\xi)|\mathbf{E}_{m \in B} e(\xi B)|^2$$

and so

$$\|f - g\|_{u^2(\mathbf{Z}/N\mathbf{Z})} = \sup_{\xi \in \mathbf{Z}/N\mathbf{Z}} |\hat{f}(\xi)|(1 - |\mathbf{E}_{m \in B} e(\xi B)|^2).$$

The frequencies ξ that lie outside ξ give a contribution of at most ε by the definition of S, so now we look at the terms where $\xi \in S$. From the definition of B and the triangle inequality we have

$$|\mathbf{E}_{m \in B} e(\xi B) - 1| \leq \varepsilon$$

in such cases, while from the measure nature of ν we have

$$|\hat{f}(\xi)| \leq \mathbf{E}_{n \in \mathbf{Z}/N\mathbf{Z}} \nu(n) = 1 + o_{N \to \infty}(1).$$

Putting this all together, we obtain

$$\|f - g\|_{u^2(\mathbf{Z}/N\mathbf{Z})} \ll \varepsilon + o_{N \to \infty}(1).$$

1.7. Linear equations in primes

To summarise, we have the following result, which essentially appears in [**GrTa2006**]:

Theorem 1.7.5 (Criterion for Roth-pseudorandomness). *Let ν be a measure obeying the Fourier-pseudorandomness assumption (1.55) and the restriction estimate (1.54) for some $2 < q < 3$. Then ν is Roth-pseudorandom.*

This turns out to be a fairly tractable criterion for establishing the Roth-pseudorandomness of various measures, which in turn can be used to detect progressions of length three (and related patterns) on various sparse sets, such as the primes; see the next section.

The above arguments to establish Roth-pseudorandomness relied heavily on linear Fourier analysis. Now we give an alternate approach that avoids Fourier analysis entirely; it is less efficient and a bit messier, but will extend in a fairly straightforward (but notationally intensive) manner to higher order patterns. To do this, we replace the u^2 norm in Lemma 1.7.3 with the U^2 norm, so now we have to verify a control by U^2 hypothesis and an approximation by U^2 hypothesis.

We begin with the control by U^2 hypothesis. Instead of Fourier analysis, we will rely solely on the Cauchy-Schwarz inequality, using a weighted version of the arguments from Section 1.3 that first appeared in [**GrTa2008b**]. We wish to control the expression

$$\Lambda(f, g, h) = \mathbf{E}_{n, r \in \mathbf{Z}/N\mathbf{Z}} f(n) g(n+r) h(n+2r)$$

where f, g, h are bounded in magnitude by $\nu + 1$. For simplicity we will just assume that f, g, h are bounded in magnitude by ν; the more general case is similar but a little bit messier. For brevity we will also omit the domain $\mathbf{Z}/N\mathbf{Z}$ in the averages, and also abbreviate $o_{N \to \infty}(1)$ as $o(1)$. We make the change of variables $(n, r) = (b + 2c, -a - b - c)$ to write this expression as

$$\mathbf{E}_{a,b,c} f(b+2c) g(a-c) h(-2a-b),$$

the point being that each term involves only two of the three variables a, b, c.

We can pointwise bound h by ν and estimate the above expression in magnitude by

$$\mathbf{E}_{a,b} |\mathbf{E}_c f(b+2c) g(a-c)| \nu(-2a-b).$$

Since $\mathbf{E}\nu = 1 + o(1)$, we can use Cauchy-Schwarz and bound this by

$$(1 + o(1))(\mathbf{E}_{a,b} |\mathbf{E}_c f(b+2c) g(a-c)|^2 \nu(-2a-b))^{1/2}$$

which we rewrite as

$$(1 + o(1)) \left(\mathbf{E}_{a,b,c,c'} f(b+2c) f(b+2c') g(a-c) g(a-c') \nu(-2a-b) \right)^{1/2}.$$

We now bound g by ν, to obtain

$$(1 + o(1)) \left(\mathbf{E}_{a,c,c'} \nu(a-c) \nu(a-c') |\mathbf{E}_b f(b+2c) f(b+2c') \nu(-2a-b)| \right)^{1/2}.$$

If we make the hypothesis

(1.56) $$\mathbf{E}_{a,c,c'}\nu(a-c)\nu(a-c') = 1 + o(1)$$

(which is a variant of (1.55), as can be seen by expanding out using Fourier analysis), followed by Cauchy-Schwarz, we can bound this by

$$(1+o(1))\left(\mathbf{E}_{a,c,c'}\nu(a-c)\nu(a-c')|\mathbf{E}_b f(b+2c)f(b+2c')\nu(-2a-b)|^2\right)^{1/4}.$$

We expand this out as

$$(1+o(1))|\mathbf{E}_{a,b,b',c,c'} f(b+2c)f(b'+2c)f(b+2c')f(b'+2c')F(b,b',c,c')|^{1/4}$$

where

$$F(b,b',c,c') := \mathbf{E}_a \nu(a-c)\nu(a-c')\nu(-2a-b)\nu(-2a-b').$$

If the F factor could be replaced by 1, then the expression inside the absolute values would just be $\|f\|_{U^2(\mathbf{Z}/N\mathbf{Z})}^4$, which is what we wanted. Applying the triangle inequality and bounding f by ν, we can thus bound the previous expression by

$$\ll \left(\|f\|_{U^2(\mathbf{Z}/N\mathbf{Z})} + \mathbf{E}_{a,b,b',c,c'}\nu(b+2c)\nu(b'+2c)\nu(b+2c')\nu(b'+2c')\right.$$
$$\left.|F(b,b',c,c')-1|\right)^{1/4}.$$

If we make the hypotheses

(1.57) $\mathbf{E}_{a,b,b',c,c'}\nu(b+2c)\nu(b'+2c)\nu(b+2c')\nu(b'+2c')F(b,b',c,c')^i = 1+o(1)$

for $i = 0, 1, 2$, then another application of Cauchy-Schwarz gives

$$\mathbf{E}_{a,b,b',c,c'}\nu(b+2c)\nu(b'+2c)\nu(b+2c')\nu(b'+2c')|F(b,b',c,c')-1| = o(1),$$

so we have obtained the control in U^2 hypothesis (at least for f, and assuming boundedness by ν and $\nu+1$ assuming the conditions (1.56), (1.57)). We refer to such conditions (involving the product of ν evaluated at distinct linear forms on the left-hand side, and a $1+o(1)$ on the right-hand side) as *linear forms conditions*. Generalising to the case of functions bounded by $\nu+1$, and permuting f, g, h, we can soon obtain the following result (stated somewhat informally):

Lemma 1.7.6 (Generalised von Neumann theorem). *If ν obeys a certain finite list of linear forms conditions, then the control by U^2 hypothesis in Lemma 1.7.3 holds.*

Now we turn to the approximation in U^2 property. It is possible to establish this approximation property by an energy increment method, analogous to the energy increment proof of Roth's theorem in Section 1.2; see [**GrTa2006**] for details. However, this argument turns out to be rather complicated. We give here a simpler approach based on duality (and more

1.7. Linear equations in primes

precisely, the Hahn-Banach theorem) that yields the same result, due independently to Gowers [**Go2010**] and to Reingold-Trevisan-Tulsiani-Vadhan [**ReTrTuVa2008**]. This approach also has the benefit of giving somewhat sharper quantitative refinements.

The first task is to represent the U^2 norm in a dual formulation. The starting point is that the expression

$$\|f\|_{U^2(\mathbf{Z}/N\mathbf{Z})}^4 = \mathbf{E}_{n,a,b} f(n) f(n+a) f(n+b) f(n+a+b)$$

whenever $f\colon \mathbf{Z}/N\mathbf{Z} \to \mathbf{R}$, can be rewritten as

$$\|f\|_{U^2(\mathbf{Z}/N\mathbf{Z})}^4 = \langle f, \mathcal{D}f \rangle_{L^2(\mathbf{Z}/N\mathbf{Z})}$$

where the *dual function* $\mathcal{D}f = \mathcal{D}_2 f\colon \mathbf{Z}/N\mathbf{Z} \to \mathbf{R}$ is defined by

$$\mathcal{D}f(n) := \mathbf{E}_{a,b} f(n+a) f(n+b) f(n+a+b).$$

Define a *basic anti-uniform function* to be any function of the form $\mathcal{D}F$, where $F\colon \mathbf{Z}/N\mathbf{Z} \to \mathbf{R}$ obeys the pointwise bound $|F| \leq \nu + 1$. To obtain the approximation property, it thus suffices to show that for every $\varepsilon > 0$, for N sufficiently large depending on ε, and any $f\colon \mathbf{Z}/N\mathbf{Z} \to \mathbf{R}$ with $0 \leq f \leq \nu$, one can decompose $f = f_1 + f_2$ where $0 \leq f_1 \leq 1$ and $|\langle f_2, \mathcal{D}F \rangle| \leq \varepsilon^4$ for all basic anti-uniform functions $\mathcal{D}F$. Indeed, if one sets $F := f_2$, the latter bound gives $\|f_2\|_{U^2(\mathbf{Z}/N\mathbf{Z})}^4 \leq \varepsilon^4$, and the desired decomposition follows.

In order to apply the Hahn-Banach theorem properly, it is convenient to symmetrise and convexify the space of basic anti-uniform functions. Define an *averaged anti-uniform function* to be any convex combination of basic anti-uniform functions and their negations, and denote the space of all such averaged anti-uniform functions as B. Thus B is a compact convex symmetric subset of the finite-dimensional real vector space $L^2(\mathbf{Z}/N\mathbf{Z})$ that contains a neighbourhood of the origin; equivalently, it defines a norm on $L^2(\mathbf{Z}/N\mathbf{Z})$. Our task is then to show (for fixed ε and large N) that for any $f \in \mathbf{Z}/N\mathbf{Z} \to \mathbf{R}$ with $0 \leq f \leq \nu + 1$, the sets

$$U := \{(f_1, f_2) \in L^2(\mathbf{Z}/N\mathbf{Z}) \cap L^2(\mathbf{Z}/N\mathbf{Z}) : f_1 + f_2 = f\}$$

and

$$V := \{(f_1, f_2) \in L^2(\mathbf{Z}/N\mathbf{Z}) \cap L^2(\mathbf{Z}/N\mathbf{Z}) : 0 \leq f_1 \leq 1; \langle f_2, \phi \rangle \leq \varepsilon^4$$
$$\text{for all } \phi \in B\}$$

have non-empty intersection.

The point of phrasing things this way is that U and V are both closed convex subsets of the finite-dimensional vector space $L^2(\mathbf{Z}/N\mathbf{Z}) \cap L^2(\mathbf{Z}/N\mathbf{Z})$, and so the *Hahn-Banach theorem* is applicable[23]. Indeed, suppose that there

---

[23] One could also use closely related results, such as the *Farkas lemma*; see [**Ta2008**, §1.16] for more discussion.

was some f for which U and V were disjoint. Then, by the Hahn-Banach theorem, there must exist some linear functional

$$(f_1, f_2) \mapsto \langle f_1, \phi_1 \rangle_{L^2(\mathbf{Z}/N\mathbf{Z})} + \langle f_2, \phi_2 \rangle_{L^2(\mathbf{Z}/N\mathbf{Z})}$$

which separates the two sets, in the sense that

$$\langle f_1, \phi_1 \rangle_{L^2(\mathbf{Z}/N\mathbf{Z})} + \langle f_2, \phi_2 \rangle_{L^2(\mathbf{Z}/N\mathbf{Z})} > c$$

for all $(f_1, f_2) \in U$, and

$$\langle f_1, \phi_1 \rangle_{L^2(\mathbf{Z}/N\mathbf{Z})} + \langle f_2, \phi_2 \rangle_{L^2(\mathbf{Z}/N\mathbf{Z})} \leq c$$

for all $(f_1, f_2) \in V$, where c is a real number.

From the form of U, we see that we must have $\phi_1 = \phi_2$. In particular, we may normalise $\phi = \phi_1 = \phi_2$ to be on the boundary of B. As all finite-dimensional spaces are reflexive, we see in that case that $\langle f_2, \phi \rangle$ can be as large as ε^4 on V, and independently $\langle f_1, \phi \rangle$ can be as large as $\mathbf{E}_{n \in \mathbf{Z}/N\mathbf{Z}} \max(\phi, 0)$. We conclude that

$$\mathbf{E}_{n \in \mathbf{Z}/N\mathbf{Z}} \max(\phi, 0) + \varepsilon^4 \leq \mathbf{E}_{n \in \mathbf{Z}/N\mathbf{Z}} f \phi.$$

As $0 \leq f \leq \nu$, we see that $f\phi \leq \nu \max(\phi, 0)$, and thus

$$\mathbf{E}_{n \in \mathbf{Z}/N\mathbf{Z}} (\nu - 1) \max(\phi, 0) \geq \varepsilon^4.$$

We now make the hypothesis that the dual function $\mathcal{D}(\nu + 1)$ of $\nu + 1$ is uniformly bounded:

(1.58) $$\mathcal{D}(\nu + 1) \leq C.$$

We remark that the linear forms condition (which we have not specified explicitly) will give this bound with $C = \mathcal{D}(1 + 1) + o(1) = 2^{2^2 - 1} + o(1)$.

Since ϕ is a convex combination of functions of the form $\pm \mathcal{D}F$ and $|F| \leq \nu + 1$, this implies that ϕ is bounded uniformly as well: $|\phi| \leq C$. Applying the Weierstrass approximation theorem to the function $\max(x, 0)$ for $|x| \leq C$ (and noting that the L^1 norm of $\nu - 1$ is $O(1)$) we conclude that there exists a polynomial $P \colon \mathbf{R} \to \mathbf{R}$ (depending only on ε and C) such that

$$\mathbf{E}_{n \in \mathbf{Z}/N\mathbf{Z}} (\nu - 1) P(\phi) \geq \varepsilon^4 / 2$$

(say). Breaking P into monomials, and using the pigeonhole principle, we conclude that there exists a non-negative integer $k = O_{\varepsilon, C}(1)$ such that

$$|\mathbf{E}_{n \in \mathbf{Z}/N\mathbf{Z}} (\nu - 1) \phi^k| \gg_{\varepsilon, C} 1;$$

since ϕ was a convex combination of functions of the form $\pm \mathcal{D}F$, we thus conclude that there exist $F_1, \ldots, F_k$ with $|F_1|, \ldots, |F_k| \leq \nu + 1$ such that

$$|\mathbf{E}_{n \in \mathbf{Z}/N\mathbf{Z}} (\nu - 1)(\mathcal{D}F_1) \ldots (\mathcal{D}F_k)| \gg_{\varepsilon, C} 1.$$

We shall contradict this by *fiat*, making the hypothesis that

(1.59) $$\mathbf{E}_{n \in \mathbf{Z}/N\mathbf{Z}} (\nu - 1)(\mathcal{D}F_1) \ldots (\mathcal{D}F_k) = o_{N \to \infty; k}(1)$$

1.7. Linear equations in primes

for all $k \geq 1$ and all $F_1, \ldots, F_k$ bounded in magnitude by $\nu + 1$.

We summarise this discussion as follows:

Theorem 1.7.7 (Dense model theorem). *If (1.58), (1.59) hold, then the approximation in U^2 hypothesis in Lemma 1.7.3 holds.*

There is nothing too special about the U^2 norm here; one could work with higher Gowers norms, or indeed with any other norm for which one has a reasonably explicit description of the dual.

The abstract version of the theorem was first (implicitly) proven in [**GrTa2008b**], and made more explicit in [**TaZi2008**]. The methods there were different (and somewhat more complicated). To prove approximation, the basic idea was to write $g = \mathbf{E}(f|\mathcal{B})$ for some carefully chosen σ-algebra $\mathcal{B}$ (built out of dual functions that correlated with things like the residual $f - \mathbf{E}(f|\mathcal{B})$). This automatically gave the non-negativity of g; the upper bound on g came from the bound $\mathbf{E}(f|\mathcal{B}) \leq \mathbf{E}(\nu|\mathcal{B})$, with the latter expression then being bounded by the Weierstrass approximation theorem and (1.59).

To summarise, in order to establish the Roth-pseudorandomness of a measure μ, we have at least two options. The first (which relies on Fourier analysis, and is thus largely restricted to complexity 1 problems) is to establish the Fourier pseudorandomness bound (1.55) and the restriction estimate (1.54). The other (which does not require Fourier analysis) is to establish a finite number of linear forms conditions, as well as the estimate (1.59).

Next, we informally sketch how one can deduce (1.59) from a finite number of linear forms conditions, as well as a crude estimate

$$(1.60) \qquad \nu = O(N^{o(1)})$$

and a condition known as the *correlation condition*. At the cost of oversimplifying slightly, we express this condition as the assertion that

$$(1.61) \qquad \mathbf{E}_{n \in \mathbf{Z}/N\mathbf{Z}} \nu(n+h_1) \ldots \nu(n+h_k) \ll_k 1$$

whenever $h_1, \ldots, h_k \in \mathbf{Z}/N\mathbf{Z}$ are distinct, thus the k-point correlation function of ν is bounded for each k. For the number-theoretic applications, one needs to replace the 1 on the right-hand side by a more complicated expression, but we will defer this technicality to the exercises. We remark that for each fixed k, the correlation condition would be implied by the linear forms condition, but it is important that we can make k arbitrarily large.

For simplicity of notation we assume that the F_j are bounded in magnitude by ν rather than by $\nu + 1$. We begin by expanding out (1.59) as

$$|\mathbf{E}_{n,h_{1,1},\ldots,h_{2,k}}(\nu(n) - 1) \prod_{j=1}^{k} F_j(n+h_{1,j}) F_j(n+h_{2,j}) F_j(n+h_{1,j}+h_{2,j})|.$$

Shifting $h_{i,j}$ by h_i for some h_1, h_2 and reaveraging, we can rewrite this as

$$|\mathbf{E}_{h_{1,1},\ldots,h_{2,k}} \mathbf{E}_{n,h_1,h_2}(\nu(n) - 1) F_{\vec{h}_1}(n+h_1) F_{\vec{h}_2}(n+h_2) F_{\vec{h}_1 + \vec{h}_2}(n+h_1+h_2)|$$

where $\vec{h}_i := (h_{i,1}, \ldots, h_{i,k})$ for $i = 1, 2$ and

$$F_{(v_1,\ldots,v_k)}(n) := \prod_{j=1}^{k} F_j(n+v_j).$$

The inner expectation is the Gowers inner product of $\nu - 1$, $F_{\vec{h}_1}$, $F_{\vec{h}_2}$, and $F_{\vec{h}_1 + \vec{h}_2}$. Using the linear forms condition we may assume that

$$\|\nu - 1\|_{U^2(\mathbf{Z}/N\mathbf{Z})} = o(1)$$

and so it will suffice by the Cauchy-Schwarz-Gowers inequality, followed by the Hölder inequality, to show that

$$\mathbf{E}_{h_{1,1},\ldots,h_{2,k}} \|F_{\vec{h}_1}\|^4_{U^2(\mathbf{Z}/N\mathbf{Z})} \ll_K 1$$

and similarly for $\vec{h}_2$ and $\vec{h}_1 + \vec{h}_2$.

We just prove the claim for $\vec{h}_1$, as the other two cases are similar. We expand the left-hand side as

$$|\mathbf{E}_{n,a,b,h_1,\ldots,h_k} \prod_{j=1}^{k} F_j(n+h_j) F_j(n+h_j+a) F_j(n+h_j+b) F_j(n+h_j+a+b)|$$

which we can upper bound by

$$|\mathbf{E}_{n,a,b,h_1,\ldots,h_k} \prod_{j=1}^{k} \nu(n+h_j) \nu(n+h_j+a) \nu(n+h_j+b) \nu(n+h_j+a+b)|.$$

We can factorise this as

$$\mathbf{E}_{a,b} |\mathbf{E}_n \nu(n) \nu(n+a) \nu(n+b) \nu(n+a+b)|^k.$$

Using (1.61), we see that the inner expectation is $O_k(1)$ as long as $0, a, b, a+b$ are distinct; in all other cases they are $O(N^{o(1)})$, by (1.60). Combining these two cases we obtain the claim.

Exercise 1.7.4. Show that (1.59) also follows from a finite number of linear forms conditions and (1.61), if the F_j are only assumed to be bounded in magnitude by $\nu + 1$ rather than ν, and the right-hand side of (1.61) is weakened to $\sum_{1 \le i < j \le m} \tau(h_i - h_j)$, where $\tau : \mathbf{Z}/N\mathbf{Z} \to \mathbf{R}^+$ is a function obeying the moment bounds $\mathbf{E}_{n \in \mathbf{Z}/N\mathbf{Z}} \tau(n)^q \ll_q 1$ for each $q \ge 1$.

The above machinery was geared to getting Roth-type lower bounds on $\Lambda(f, f, f)$; but it also can be used to give more precise asymptotics:

1.7. Linear equations in primes

Exercise 1.7.5. Suppose that ν obeys the hypotheses of Lemma 1.7.3 (with the u^2 norm). Let $f\colon \mathbf{Z}/N\mathbf{Z} \to \mathbf{R}$ obey the pointwise bound $0 \le f \le 1$ and has mean $\mathbf{E}_{n\in\mathbf{Z}/N\mathbf{Z}} f(n) = \delta$; suppose also that one has the pseudo-randomness bound $\sup_{\xi\in\mathbf{Z}/N\mathbf{Z}\setminus 0} |\hat f(\xi)| = o_{N\to\infty}(1)$. Show that $\Lambda(f,f,f) = \delta^3 + o_{N\to\infty}(1)$.

Exercise 1.7.6. Repeat the previous exercise, but with the u^2 norm replaced by the U^2 norm.

Informally, the above exercises show that if one wants to obtain asymptotics for three-term progressions in a set A which has positive relative density with respect to a Roth-pseudorandom measure, then it suffices to obtain a non-trivial bound on the exponential sums $\sum_{n\in A} e(\xi n)$ for non-zero frequencies ξ.

For longer progressions, one uses higher-order Gowers norms, and a similar argument (using the inverse conjecture for the Gowers norms) shows (roughly speaking) that to obtain asymptotics for k-term progressions (or more generally, linear patterns of complexity $k-1$) in a U^{k-1}-pseudorandom measure (by which we mean that the analogue of Lemma 1.7.3 for the U^{k-1} norm holds) then it suffices to obtain a non-trivial bound on sums of the form $\sum_{n\in A} F(g(n)\Gamma)$ for $k-2$-step nilsequences $F(g(n)\Gamma)$. See [**GrTa2010**] for further discussion.

1.7.2. A brief discussion of sieve theory.
In order to apply the above theory to find patterns in the primes, we need to build a measure ν with respect to which the primes have a positive density, and for which one can verify conditions such as the Fourier pseudorandomness condition (1.55), the restriction estimate (1.54), linear forms conditions, and the correlation condition (1.61).

There is an initial problem with this, namely that the primes themselves are not uniformly distributed with respect to small moduli. For instance, all primes are coprime to two (with one exception). In contrast, any measure ν obeying the Fourier pseudorandomness condition (1.55) (which is implied by the condition $\|\nu - 1\|_{U^2} = o(1)$, which would follow in turn from the linear forms condition), must be evenly distributed in both odd and even residue classes up to $o(1)$ errors; this forces the density of the primes in ν to be at most $1/2 + o(1)$. A similar argument using all the prime moduli less than some parameter w shows in fact that the density of primes in ν is at most $\prod_{p<w}(1-\frac{1}{p}) + o_{N\to\infty;w}(1)$. Since $\sum_p \frac{1}{p}$ diverges to $+\infty$, $\prod_p(1-\frac{1}{p})$ diverges to zero, and so we see that the primes cannot in fact have a positive density with respect to any pseudorandom measure.

This difficulty can be overcome by a simple affine change of variables known as the *W-trick*, where we replace the primes $\mathcal{P} = \{2,3,5,\dots\}$ by the modified set $\mathcal{P}_{W,b} := \{n \in \mathbf{N} : Wn + b \in \mathcal{P}\}$, where $W := \prod_{p<w} p$ is the product of all the primes less than w, and $1 \leq b < W$ is a residue class coprime to W. In practice, w (and W) are slowly growing functions of N, e.g., one could take $w = \log \log \log N$. By the pigeonhole principle, for any given N and W there will exist a b for which $\mathcal{P}_{W,b}$ is large (of cardinality $\gg \frac{N}{\phi(W) \log N}$, where $\phi(W)$ is the number of residue classes coprime to W); indeed, thanks to the prime number theorem in arithmetic progressions, any such b would work (e.g. one can take $b = 1$). Note that every arithmetic progression in $\mathcal{P}_{W,b}$ is associated to a corresponding arithmetic progression in $\mathcal{P}$. Thus, for the task of locating arithmetic progressions at least, we may as well work with $\mathcal{P}_{W,b}$; a similar claim also holds for more complicated tasks, such as counting the number of linear patterns in $\mathcal{P}$, though one now has to work with several residue classes at once. The point of passing from $\mathcal{P}$ to $\mathcal{P}_{W,b}$ is that the latter set no longer has any particular bias to favour or disfavour any residue class with modulus less than w; there are still biases at higher moduli, but as long as w goes to infinity with N, the effect of such biases will end up being negligible (ultimately contributing $o(1)$ terms to things like the linear forms condition).

To simplify the exposition a bit, though, let us ignore the W-trick and pretend that we are working with the primes themselves rather than the affine-shifted primes. We will also ignore the technical distinctions between the interval $[N]$ and the cyclic group $\mathbf{Z}/N\mathbf{Z}$.

The most natural candidate for the measure ν is the *von Mangoldt function* $\Lambda \colon \mathbf{N} \to \mathbf{R}^+$, defined by setting $\Lambda(n) := \log p$ when $n = p^j$ is a prime p or a power of a prime, and $\Lambda(n) = 0$ otherwise. One hint as to the significance of this function is provided by the identity

$$\log n = \sum_{d|n} \Lambda(d)$$

for all natural numbers n, which can be viewed as a generating function of the fundamental theorem of arithmetic.

The prime number theorem tells us that Λ is indeed a measure:

$$\mathbf{E}_{n \in [N]} \Lambda(n) = 1 + o(1),$$

and the primes have full density with respect to this function:

$$\mathbf{E}_{n \in [N]} 1_\mathcal{P}(n) \Lambda(n) = 1 + o(1).$$

Furthermore, the von Mangoldt function has good Fourier pseudorandomness properties (after applying the W-trick), thanks to the classical techniques of Hardy-Littlewood and Vinogradov. Indeed, to control exponential sums such as $\mathbf{E}_{n \in [N]} \Lambda(n) e(\xi n)$ for some $\xi \in \mathbf{R}$, one can use tools such as the Siegel-Walfisz theorem (a quantitative version of the prime number theorem

1.7. Linear equations in primes

in arithmetic progressions) to control such sums in the "major arc" case when ξ is close to a rational of small height, while in the "minor arc" case when ξ behaves irrationally, one can use the standard identity

$$(1.62) \qquad \Lambda(n) = \sum_{d|n} \mu(d) \log \frac{n}{d},$$

where μ is the *Möbius function*[24], to re-express such a sum in terms of expressions roughly of the form

$$\sum_{d,m} \mu(d) \log m \, e(\xi dm)$$

where we are intentionally vague as to what range the d, m parameters are being summed over. The idea is then to eliminate the μ factor by tools such as the triangle inequality or the Cauchy-Schwarz inequality, leading to expressions such as

$$\sum_{d} |\sum_{m} \log m \, e(\xi dm)|;$$

the point is that the inner sum does not contain any number-theoretic factors such as Λ or μ, but is still oscillatory (at least if ξ is sufficiently irrational), and so one can extract useful cancellation from here. Actually, the situation is more complicated than this, because there are regions of the range of (d, m) for which this method provides insufficient cancellation, in which case one has to rearrange the sum further using more arithmetic identities such as (1.62) (for instance, using a truncated version of (1.62) known as *Vaughan's identity*). We will not discuss this further here, but any advanced analytic number theory text (e.g. [**IwKo2004**]) will cover this material.

Unfortunately, while the Fourier-pseudorandomness of Λ is well understood, the linear forms and correlation conditions are essentially equivalent to (and in fact slightly harder than) the original problem of obtaining asymptotics for linear patterns in primes, and so using Λ for the pseudorandom measure would result in a circular argument. Furthermore, correlations such as

$$\mathbf{E}_{n \in [N]} \Lambda(n) \Lambda(n+2)$$

(which essentially counts the number of twin primes up to N) are notoriously difficult to compute. For instance, if one tries to expand the above sum using (1.62), one ends up with expressions such as

$$\sum_{d,d' \leq N} \mu(d)\mu(d') \sum_{n \leq N : d|n, d'|n+2} \log \frac{n}{d} \log \frac{n+2}{d'}.$$

---

[24] The *Möbius function* μ is defined by setting $\mu(n) := (-1)^k$ when n is the product of k distinct primes for some $k \geq 0$, and $\mu(n) = 0$ otherwise.

By the Chinese remainder theorem, the two residue conditions $d|n$ and $d'|n+2$ can be combined to a single residue condition for n modulo the least common multiple $lcm(d,d')$ of d and d'. If d and d' are both small, e.g., $d, d' \leq N^{1/10}$, then this least common multiple is much less than N, and in such a case one can compute the inner sum very precisely; as it turns out, the main term in this estimate is multiplicative in d, d', which allows the outer sum to be estimated using the techniques of multiplicative number theory (and in particular, using the theory of the Riemann zeta function). Unfortunately, for the bulk of the above sum, d and d' are instead comparable to N, and the least common multiple is typically of size N^2, and then it becomes extraordinarily difficult to estimate the inner sum (and hence the entire sum).

However, if we *truncate* the divisor sum (1.62) to restrict d to a range such as $d \leq N^{1/10}$, then the situation improves substantially. This leads to expressions such as

$$(1.63) \qquad \nu(n) := \frac{1}{\log R} \left(\sum_{d|n; d<R} \mu(d) \log \frac{R}{d} \right)^2$$

or, more generally,

$$(1.64) \qquad \nu(n) := \log R \left(\sum_{d|n} \mu(d) \psi\left(\frac{\log d}{\log R} \right) \right)^2$$

for some cutoff function ψ, where R is a small power[25] of N; the expression (1.63) corresponds to the case $\psi(x) := \max(1-x, 0)$. The presence of the square is to ensure that ν is non-negative, and the presence of the $\frac{1}{\log R}$ is a normalisation factor to ensure that ν has mean close to 1. Such expressions were essentially introduced to Selberg (as part of what is now known as the *Selberg sieve*), although the sieve weight factors $\psi(\frac{\log d}{\log R})$ are usually modified slightly for the Selberg sieve (see [**GrTa2006**] for further discussion). The correlation properties of the particular expression (1.63) were studied intensively by Goldston and Yıldırım (see e.g. [**GoPiYi2008**]), and have particularly sharp estimates, although for applications discussed here, one can work instead with a smoother choice of cutoff ψ, which makes the required correlation estimates on ν easier to prove (but with slightly worse bounds). Indeed, the required linear forms and correlation conditions can be verified for (1.64) (or more precisely, a variant of ν in which the W-trick

---

[25]The exact power of N that one sets R equal to will depend on the complexity of the linear forms and correlation conditions one needs. For counting progressions of length three, for instance, one can take $R = N^{1/10}$.

1.7. Linear equations in primes

is applied) by a moderately lengthy, but elementary and straightforward calculation, based ultimately on the Chinese remainder theorem, an analysis of the local problem (working mod q for small q), and the fundamental fact that the Riemann zeta function $\zeta(s)$ is approximately equal to $1/(s-1)$ for s close to 1. See for instance [**Ta2004**] for more discussion.

If one uses (1.63), then we see that $\nu(n)$ is equal to $\log R$ when n is any prime larger than R; if $\log R$ is comparable to $\log N$, we thus see (from the prime number theorem) that the primes in $[N]$ do indeed have positive density relative to ν. This is then enough to be able to invoke the transference principle and extend results such as Szemerédi's theorem to the primes, establishing, in particular, that the primes contain arbitrarily long arithmetic progressions; see [**GrTa2008b**] for details.

To use the Fourier-analytic approach, it turns out to be convenient to replace the above measures ν by a slight variant which looks more complicated in the spatial domain, but is easier to manipulate in the frequency domain. More specifically, the expression (1.63) or (1.64) is replaced with a variant such as

$$\nu := \log R \left(\sum_{d|n; d \leq R} \mu(d) \frac{d}{\phi(d)} \sum_{q \leq R/d; (q,d)=1} \frac{1}{\phi(q)} \right)^2$$

where $\phi(d)$ is the Euler totient function (the number of integers from 1 to d that are coprime to d). Some standard multiplicative number theory shows that the weights $\frac{d}{\phi(d)} \sum_{q \leq R/d; (q,d)=1} \frac{1}{\phi(q)}$ are approximately equal to $\log \frac{R}{d}$ in some sense. With such a representation, it turns out that the Fourier coefficients of ν can be computed more or less explicitly, and is essentially supported on those frequencies of the form a/q with $q \leq R^2$. This makes it easy to verify the required Fourier-pseudorandomness hypothesis (1.55) (once one applies the W-trick). As for the restriction estimate (1.54), the first step is to use Exercise (1.7.3) and the Cauchy-Schwarz inequality to reduce matters to showing an estimate of the shape

$$\mathbf{E}_n |\sum_\xi g(\xi) e(\xi nx/N)|^2 \nu(n) \ll \|g\|_{\ell^{q'}}.$$

The right-hand side can be rearranged to be of the shape

$$\sum_{\xi, \xi'} g(\xi) \overline{g(\xi')} \hat{\nu}(\xi - \xi').$$

It is then possible to use the good pointwise control on the Fourier transform $\hat{\nu}$ of ν (in particular, the fact that it "decays" quite rapidly away from the major arcs) to get a good restriction estimate. See [**GrTa2006**] for further discussion.

As discussed in the previous section, to get asymptotics for patterns in the primes we also need to control exponential sums such as
$$\sum_{p \leq N} e(\xi p)$$
and more generally (for higher complexity patterns)
$$\sum_{p \leq N} F(g(p)\Gamma)$$
for various nilsequences $n \mapsto F(g(n)\Gamma)$. Again, it is convenient to use the von Mangoldt function Λ as a proxy for the primes, thus leading to expressions such as
$$\sum_{n \leq N} \Lambda(n) F(g(n)\Gamma).$$
Actually, for technical reasons it is convenient to use identities such as (1.62) to replace this type of expression with expressions such as
$$\sum_{n \leq N} \mu(n) F(g(n)\Gamma),$$
because the Möbius function μ enjoys better boundedness and equidistribution properties than Λ. (For instance, Λ strongly favours odd numbers over even numbers, whereas the Möbius function has no preference.) It turns out that these expressions can be controlled by a generalisation of the method of Vinogradov used to compute exponential sums over primes, using the equidistribution theory of nilsequences as a substitute for the classical theory of exponential sums over integers. See [**GrTa2008c**] for details.

Chapter 2

Related articles

2.1. Ultralimit analysis and quantitative algebraic geometry

There is a close relationship between finitary (or "hard", or "quantitative") analysis, and infinitary (or "soft", or "qualitative") analysis; see e.g. [**Ta2008**, §1.3, 1.5] or [**Ta2010b**, §2.11]. One way to connect the two types of analysis is via *compactness arguments* (and more specifically, *contradiction and compactness* arguments); such arguments can convert qualitative properties (such as continuity) to quantitative properties (such as bounded), basically because of the fundamental fact that continuous functions on a compact space are bounded (or the closely related fact that sequentially continuous functions on a sequentially compact space are bounded).

A key stage in any such compactness argument is the following: one has a sequence X_n of "quantitative" or "finitary" objects or spaces, and one has to somehow end up with a "qualitative" or "infinitary" limit object X or limit space. One common way to achieve this is to embed everything inside some universal space and then use some weak compactness property of that space, such as the *Banach-Alaoglu theorem* (or its sequential counterpart; see [**Ta2010**, §1.8]). This is, for instance, the idea behind the *Furstenberg correspondence principle* relating ergodic theory to combinatorics; see for instance [**Ta2009**, §2.10] for further discussion.

However, there is a slightly different approach, which I will call *ultralimit analysis*, which proceeds via the machinery of *ultrafilters* and *ultraproducts*; typically, the limit objects X one constructs are now the ultraproducts (or ultralimits) of the original objects X_α. There are two main facts that make ultralimit analysis powerful. The first is that one can take ultralimits of *arbitrary* sequences of objects, as opposed to more traditional tools such as metric completions, which only allow one to take limits of *Cauchy sequences* of objects. The second fact is *Los's theorem*, which tells us that X is an *elementary limit* of the X_α (i.e., every sentence in first-order logic which is true for the X_α for α large enough, is true for X). This existence of elementary limits is a manifestation of the *compactness theorem* in logic; see [**Ta2010b**, §1.4] for more discussion. So we see that compactness methods and ultrafilter methods are closely intertwined[1].

Ultralimit analysis is very closely related to *non-standard analysis*; see [**Ta2008**, §1.5] for further discussion. We will expand upon this connection later in this section. Roughly speaking, the relationship between ultralimit analysis and non-standard analysis is analogous to the relationship between measure theory and probability theory.

To illustrate how ultralimit analysis is actually used in practice, we will take here a qualitative infinitary theory—in this case, basic algebraic

---

[1] See also [**Ta2010**, §1.8] for a related connection between ultrafilters and compactness.

2.1. Ultralimit analysis

geometry—and apply ultralimit analysis to then deduce a quantitative version of this theory, in which the complexity of the various algebraic sets and varieties that appear as outputs are controlled uniformly by the complexity of the inputs. The point of this exercise is to show how ultralimit analysis allows for a relatively painless conversion back and forth between the quantitative and qualitative worlds, though in some cases the quantitative translation of a qualitative result (or vice versa) may be somewhat unexpected. In a recent paper [**BrGrTa2010**], ultralimit analysis was used to reduce the messiness of various quantitative arguments by replacing them with a qualitative setting in which the theory becomes significantly cleaner.

For the sake of completeness, we will also reprove some earlier instances of the correspondence principle via ultralimit analysis, namely the deduction of the quantitative Gromov theorem from the qualitative one, and of Szemerédi's theorem from the Furstenberg recurrence theorem, to illustrate how close the two techniques are to each other.

2.1.1. Ultralimit analysis. In order to perform ultralimit analysis, we need to prepare the scene by deciding on three things in advance:

(i) The *standard universe* $\mathcal{U}$ of standard objects and spaces.

(ii) A distinction between *ordinary objects*, and *spaces*.

(iii) A choice of *non-principal ultrafilter* $\alpha_\infty \in \beta \mathbf{N} \backslash \mathbf{N}$.

We now discuss each of these three preparatory ingredients in turn.

We assume that we have a *standard universe* or *superstructure* $\mathcal{U}$ which contains all the "standard" sets, objects, and structures that we ordinarily care about, such as the natural numbers, the real numbers, the power set of real numbers, the power set of the power set of real numbers, and so forth. For technical reasons, we have to limit the size of this universe by requiring that it be a set, rather than a class; thus (by *Russell's paradox*), not all sets will be standard (e.g., $\mathcal{U}$ itself will not be a standard set). However, in many areas of mathematics (particularly those of a "finitary" or at most "countable" flavour, or those based on finite-dimensional spaces such as $\mathbf{R}^d$), the type of objects considered in a field of mathematics can often be contained inside a single set $\mathcal{U}$. For instance, the class of all groups is too large to be a set. But in practice, one is only interested in, say, groups with an at most countable number of generators, and if one then enumerates these generators and considers their relations, one can identify each such group (up to isomorphism) to one in some fixed set of model groups. One can then take $\mathcal{U}$ to be the collection of these groups, and the various objects one can form from these groups (e.g., power sets, maps from one group to another, etc.). Thus, in practice, the requirement that we limit the scope of objects to care about is not a significant limitation[2].

It is important to note that while we primarily *care* about objects inside the standard universe $\mathcal{U}$, we allow ourselves to *use* objects outside the standard universe (but still inside the ambient set theory) whenever it is convenient to do so. The situation is analogous to that of using complex analysis to solve real analysis problems; one may only care about statements that have to do with real numbers, but sometimes it is convenient to introduce complex numbers within the *proofs* of such statements[3].

We will also assume that there is a distinction between two types of objects in this universe: *spaces*, which are sets that can contain other objects, and *ordinary objects*, which are all the objects that are not spaces. Thus, for instance, a group element would typically be considered an ordinary object, whereas a group itself would be a space that group elements can live in. It is also convenient to view functions $f \colon X \to Y$ between two spaces as itself being a type of ordinary object (namely, an element of a space $\mathrm{Hom}(X, Y)$ of maps from X to Y). The precise concept of what constitutes a space, and what constitutes an ordinary object, is somewhat hard to formalise, but the basic rule of thumb to decide whether an object X should be a space or not is to ask whether mathematical phrases such as $x \in X$, $f \colon X \to Y$, or $A \subset X$ are likely to make useful sense. If so, then X is a space; otherwise, X is an ordinary object.

Examples of spaces include sets, groups, rings, fields, graphs, vector spaces, topological spaces, metric spaces, function spaces, measure spaces, dynamical systems, and operator algebras. Examples of ordinary objects include points, numbers, functions, matrices, strings, and equations.

Remark 2.1.1. Note that in some cases, a single object may seem to be both an ordinary object and a space, but one can often separate the two roles that this object is playing by making a sufficiently fine distinction. For instance, in Euclidean geometry, a line ℓ in is both an ordinary object (it is one of the primitive concepts in that geometry), but it can also be viewed as a space of points. In such cases, it becomes useful to distinguish between the *abstract line* ℓ, which is the primitive object, and its *realisation* $\ell[\mathbf{R}]$ as a space of points in the Euclidean plane. This type of distinction is quite common in algebraic geometry, thus, for instance, the imaginary circle $C := \{(x, y) : x^2 + y^2 = -1\}$ has an empty realisation $C[\mathbf{R}] = \emptyset$ in the real plane $\mathbf{R}^2$, but has a non-trivial realisation $C[\mathbf{C}]$ in the complex

---

[2] If one does not want to limit one's scope in this fashion, one can proceed instead using the machinery of *Grothendieck universes*.

[3] More generally, the trick of passing to some *completion* $\overline{\mathcal{U}}$ of one's original structure $\mathcal{U}$ in order to more easily perform certain mathematical arguments is a common theme throughout modern mathematics.

2.1. Ultralimit analysis

plane $\mathbf{C}^2$ (or over finite fields), and so we do not consider C (as an abstract algebraic variety) to be empty. Similarly, given a function f, we distinguish between the function f itself (as an abstract object) and the graph $f[X] := \{(x, f(x)) : x \in X\}$ of that function over some given domain X.

We also fix a *non-principal ultrafilter* α_∞ on the natural numbers. Recall that this is a collection of subsets of $\mathbf{N}$ with the following properties:

(i) No finite set lies in α_∞.

(ii) If $A \subset \mathbf{N}$ is in α_∞, then any subset of $\mathbf{N}$ containing A is in α_∞.

(iii) If A, B lie in α_∞, then $A \cap B$ also lies in α_∞.

(iv) If $A \subset \mathbf{N}$, then exactly one of A and $\mathbf{N}\backslash A$ lies in α_∞.

Given a property $P(\alpha)$ which may be true or false for each natural number α, we say that P is true for α *sufficiently close to* α_∞ if the set $\{\alpha \in \mathbf{N} : P(\alpha) \text{ holds}\}$ lies in α_∞. The existence of a non-principal ultrafilter α_∞ is guaranteed by the *ultrafilter lemma*, which can be proven using the axiom of choice (or equivalently, by using Zorn's lemma).

Remark 2.1.2. One can view α_∞ as a point in the *Stone-Čech compactification* (see [Ta2010, §1.8]), in which case "for α sufficiently close to α_∞" acquires the familiar topological meaning "for all α in a neighbourhood of α_∞".

We can use this ultrafilter to take limits of standard objects and spaces. Indeed, given any two sequences $(x_\alpha)_{\alpha \in \mathbf{N}}$, $(y_\alpha)_{\alpha \in \mathbf{N}}$ of standard ordinary objects, we say that such sequences are *equivalent* if we have $x_\alpha = y_\alpha$ for all α sufficiently close to α_∞. We then define the *ultralimit* $\lim_{\alpha \to \alpha_\infty} x_\alpha$ of a sequence $(x_\alpha)_{\alpha \in \mathbf{N}}$ to be the equivalence class of $(x_\alpha)_{\alpha \in \mathbf{N}}$ (in the space $\mathcal{U}^{\mathbf{N}}$ of all sequences in the universe). In other words, we have

$$\lim_{\alpha \to \alpha_\infty} x_\alpha = \lim_{\alpha \to \alpha_\infty} y_\alpha$$

if and only if $x_\alpha = y_\alpha$ for all α sufficiently close to α_∞.

The ultralimit $\lim_{\alpha \to \alpha_\infty} x_\alpha$ lies outside the standard universe $\mathcal{U}$, but is still constructible as an object in the ambient set theory (because $\mathcal{U}$ was assumed to be a set). Note that we do not need x_α to be well defined for all α for the limit $(x_\alpha)_{\alpha \in \mathbf{N}}$ to make sense; it is enough that x_α is well defined for all α sufficiently close to α_∞.

If $x = \lim_{\alpha \to \alpha_\infty} x_\alpha$, we refer to the sequence x_α of ordinary objects as a *model* for the limit x. Thus, any two models for the same limit object x will agree in a sufficiently small neighbourhood of α_∞.

Similarly, given a sequence of standard spaces $(X_\alpha)_{\alpha \in \mathbf{N}}$, one can form[4] the *ultralimit* (or *ultraproduct*) $\lim_{\alpha \to \alpha_\infty} X_\alpha$, defined as the collection of all ultralimits $\lim_{\alpha \to \alpha_\infty} x_\alpha$ of sequences x_α, where $x_\alpha \in X_\alpha$ for all $\alpha \in \mathbf{N}$ (or for all α sufficiently close to α_∞). Again, this space will lie outside the standard universe, but is still a set. If $X = \lim_{\alpha \to \alpha_\infty} X_\alpha$, we refer to the sequence X_α of spaces as a *model* for X.

As a special case of an ultralimit, given a single space X, its ultralimit $\lim_{\alpha \to \alpha_\infty} X$ is known as the *ultrapower* of X and will be denoted $^*X$.

Remark 2.1.3. One can view $^*X$ as a type of *completion* of X, much as the reals are the *metric completion* of the rationals. Indeed, just as the reals encompass all limits $\lim_{n \to \infty} x_n$ of Cauchy sequences $x_1, x_2, \ldots$ in the rationals, up to equivalence, the ultrapower $^*X$ encompass all limits of *arbitrary* sequences in X, up to agreement sufficiently close to α_∞. The ability[5] to take limits of arbitrary sequences, and not merely Cauchy sequences or convergent sequences, is the underlying source of power of ultralimit analysis.

Of course, we embed the rationals into the reals by identifying each rational x with its limit $\lim_{n \to \infty} x$. In a similar spirit, we identify every standard ordinary object x with its ultralimit $\lim_{\alpha \to \alpha_\infty} x$. In particular, a standard space X is now identified with a subspace of $^*X$. When X is finite, it is easy to see that this embedding of X to $^*X$ is surjective; but for infinite X, the ultrapower is significantly larger than X itself.

Remark 2.1.4. One could collect the ultralimits of all the ordinary objects and spaces in the standard universe $\mathcal{U}$ and form a new structure, the *non-standard universe* $\overline{\mathcal{U}}_{\alpha_\infty}$, which one can view as a *completion* of the standard universe, in much the same way that the reals are a completion of the rationals. However, we will not have to explicitly deal with this non-standard universe and will not discuss it again in this book.

In non-standard analysis, an ultralimit of standard ordinary object in a given class is referred to as (or more precisely, *models*) a *non-standard* object in that class. To emphasise the slightly different philosophy of ultralimit analysis, however, I would like to call these objects *limit objects* in that class instead. Thus, for instance:

(i) An ultralimit $n = \lim_{\alpha \to \alpha_\infty} n_\alpha$ of standard natural numbers is a *limit natural number* (or a non-standard natural number, or an element of $^*\mathbf{N}$).

---

[4]This will not conflict with the notion of ultralimits for ordinary objects, so long as one always takes care to keep spaces and ordinary objects separate.

[5]This ability ultimately arises from the universal nature of the Stone-Čech compactification $\beta \mathbf{N}$, as well as the discrete nature of $\mathbf{N}$, which makes all sequences $n \mapsto x_n$ continuous.

2.1. Ultralimit analysis

(ii) An ultralimit $x = \lim_{\alpha \to \alpha_\infty} x_\alpha$ of standard real numbers is a *limit real number* (or a non-standard real number, or a *hyperreal*, or an element of *$\mathbf{R}$).

(iii) An ultralimit $\phi = \lim_{\alpha \to \alpha_\infty} \phi_\alpha$ of standard functions $\phi_\alpha \colon X_\alpha \to Y_\alpha$ between two sets X_α, Y_α is a *limit function* (also known as an *internal function*, or a *non-standard function*).

(iv) An ultralimit $\phi = \lim_{\alpha \to \alpha_\infty} \phi_\alpha$ of standard continuous functions $\phi_\alpha \colon X_\alpha \to Y_\alpha$ between two topological spaces X_α, Y_α is a *limit continuous function* (or *internal continuous function*, or *non-standard continuous function*).

(v) Etc.

Clearly, all standard ordinary objects are limit objects of the same class, but not conversely.

Similarly, ultralimits of spaces in a given class will be referred to as *limit spaces* in that class (in non-standard analysis, they would be called non-standard spaces or internal spaces instead). For instance:

(i) An ultralimit $X = \lim_{\alpha \to \alpha_\infty} X_\alpha$ of standard sets is a limit set (or internal set, or non-standard set).

(ii) An ultralimit $G = \lim_{\alpha \to \alpha_\infty} G_\alpha$ of standard groups is a limit group (or internal group, or non-standard group).

(iii) An ultralimit $(X, \mathcal{B}, \mu) = \lim_{\alpha \to \alpha_\infty} (X_\alpha, \mathcal{B}_\alpha, \mu_\alpha)$ of standard measure spaces is a limit measure space (or internal measure space, or non-standard measure space).

(iv) Etc.

Note that finite standard spaces will also be limit spaces of the same class, but infinite standard spaces will not. For instance, $\mathbf{Z}$ is a standard group, but is not a limit group, basically because it does not contain limit integers such as $\lim_{\alpha \to \alpha_\infty} \alpha$. However, $\mathbf{Z}$ is contained in the limit group *$\mathbf{Z}$. The relationship between standard spaces and limit spaces is analogous to that between incomplete spaces and complete spaces in various fields of mathematics (e.g. in metric space theory or field theory).

Any operation or result involving finitely many standard objects, spaces, and first-order quantifiers carries over to their non-standard or limit counterparts (the formal statement of this is *Los's theorem*). For instance, the addition operation on standard natural numbers gives an addition operation on limit natural numbers, defined by the formula

$$\lim_{\alpha \to \alpha_\infty} n_\alpha + \lim_{\alpha \to \alpha_\infty} m_\alpha := \lim_{\alpha \to \alpha_\infty} (n_\alpha + m_\alpha).$$

It is easy to see that this is a well-defined operation on the limit natural numbers *$\mathbf{N}$, and that the usual properties of addition (e.g. the associative and commutative laws) carry over to this limit (much as how the associativity and commutativity of addition on the rationals automatically implies the same laws of arithmetic for the reals). Similarly, we can define the other arithmetic and order relations on limit numbers; for instance, we have

$$\lim_{\alpha \to \alpha_\infty} n_\alpha \geq \lim_{\alpha \to \alpha_\infty} m_\alpha$$

if and only if $n_\alpha \geq m_\alpha$ for all α sufficiently close to α_0, and similarly define $\leq, >, <$, etc. Note from the definition of an ultrafilter that we still have the usual order trichotomy: given any two limit numbers n, m, exactly one of $n < m$, $n = m$, and $n > m$ is true.

Example 2.1.5. The limit natural number $\omega := \lim_{\alpha \to \alpha_\infty} \alpha$ is larger than all standard natural numbers, but $\omega^2 = \lim_{\alpha \to \alpha_\infty} \alpha^2$ is even larger still.

The following two exercises should give some intuition of how Łos's theorem is proved, and what it could be useful for:

Exercise 2.1.1. Show that the following two formulations of *Goldbach's conjecture* are equivalent:

(i) Every even natural number greater than two is the sum of two primes.

(ii) Every even limit natural number greater than two is the sum of two prime limit natural numbers.

Here, we define a limit natural number n to be *even* if we have $n = 2m$ for some limit natural number m, and a limit natural number n to be *prime* if it is greater than 1 but cannot be written as the product of two limit natural numbers greater than 1.

Exercise 2.1.2. Let k_α be a sequence of algebraically closed fields. Show that the ultralimit $k := \lim_{\alpha \to \alpha_\infty} k_\alpha$ is also an algebraically closed field. In other words, every limit algebraically closed field is an algebraically closed field.

Given an ultralimit $\phi := \lim_{\alpha \to \alpha_\infty} \phi_\alpha$ of functions $\phi_\alpha \colon X_\alpha \to Y_\alpha$, we can view ϕ as a function from the limit space $X := \prod_{\alpha \to \alpha_\infty} X_\alpha$ to the limit space $Y := \prod_{\alpha \to \alpha_\infty} Y_\alpha$ by the formula

$$\phi(\lim_{\alpha \to \alpha_\infty} x_\alpha) := \lim_{\alpha \to \alpha_\infty} \phi_\alpha(x_\alpha).$$

Again, it is easy to check that this is well defined. Thus every limit function from a limit space X to a limit space Y is a function from X to Y, but the converse is not true in general.

One can easily show that limit sets behave well with respect to finitely many Boolean operations; for instance, the intersection of two limit sets $X = \lim_{\alpha \to \alpha_\infty} X_\alpha$ and $Y = \lim_{\alpha \to \alpha_\infty} Y_\alpha$ is another limit set, namely $X \cap Y = \lim_{\alpha \to \alpha_\infty} X_\alpha \cap Y_\alpha$. However, we caution that the same is not necessarily true for infinite Boolean operations; the countable union or intersection of limit sets need not be a limit set. (For instance, each individual standard integer in $\mathbf{Z}$ is a limit set, but their union $\mathbf{Z}$ is not.) Indeed, there is an analogy between the limit subsets of a limit set, and the *clopen* (simultaneously closed and open) subsets of a topological space (or the *constructible sets* in an algebraic variety).

By the same type of arguments used to show Exercise 2.1.2, one can check that every limit group is a group (albeit one that usually lies outside the standard universe $\mathcal{U}$), every limit ring is a ring, every limit field is a field, etc.

The situation with vector spaces is a little more interesting. The ultraproduct $V = \lim_{\alpha \to \alpha_\infty} V_\alpha$ of a collection of standard vector spaces V_α over $\mathbf{R}$ is a vector space over the larger field $^*\mathbf{R}$, because the various scalar multiplication operations $\cdot_\alpha \colon \mathbf{R} \times V_\alpha \to V_\alpha$ over the standard reals become a scalar multiplication operation $\cdot \colon {}^*\mathbf{R} \times V \to V$ over the limit reals. Of course, as the standard reals $\mathbf{R}$ are a subfield of the limit reals $^*\mathbf{R}$, V is also a vector space over the standard reals $\mathbf{R}$; but when viewed this way, the properties of the V_α are not automatically inherited by V. For instance, if each of the V_α are d-dimensional over $\mathbf{R}$ for some fixed finite d, then V is d-dimensional over the *limit* reals $^*\mathbf{R}$, but is infinite dimensional over the reals $\mathbf{R}$.

Now let $A = \lim_{\alpha \to \alpha_\infty} A_\alpha$ be a limit finite set, i.e., a limit of finite sets A_α. Every finite set is a limit finite set, but not conversely; for instance, $\lim_{\alpha \to \alpha_\infty} \{1, \ldots, \alpha\}$ is a limit finite set which has infinite cardinality. On the other hand, because every finite set A_α has a cardinality $|A_\alpha| \in \mathbf{N}$ which is a standard natural number, we can assign to every limit finite set $A = \lim_{\alpha \to \alpha_\infty} A_\alpha$ a *limit cardinality* $|A| \in {}^*\mathbf{N}$ which is a limit natural number, by the formula

$$|\lim_{\alpha \to \alpha_\infty} A_\alpha| := \lim_{\alpha \to \alpha_\infty} |A_\alpha|.$$

This limit cardinality inherits all of the first-order properties of ordinary cardinality. For instance, we have the inclusion-exclusion formula

$$|A \cup B| + |A \cap B| = |A| + |B|$$

for any two limit finite sets; this follows from the inclusion-exclusion formula for standard finite sets by an easy limiting argument.

It is not hard to show that $\lim_{\alpha \to \alpha_\infty} A_\alpha$ is finite if and only if the $|A_\alpha|$ are bounded for α sufficiently close to α_∞. Thus, we see that one feature of passage to ultralimits is that it converts the term "bounded" to "finite", while the term "finite" becomes "limit finite". This makes ultralimit analysis useful for deducing facts about bounded quantities from facts about finite quantities. We give some examples of this in the next section.

In a similar vein, an ultralimit $(X, d) = \lim_{\alpha \to \alpha_\infty}(X_\alpha, d_\alpha)$ of standard metric spaces (X_α, d_α) yields a *limit* metric space; thus, for instance, $d \colon X \times X \to {}^*\mathbf{R}$ is now a metric taking values in the *limit* reals. Now, if the spaces (X_α, d_α) were uniformly bounded, then the limit space (X, d) would be bounded by some (standard) real diameter. From the *Bolzano-Weierstrass theorem* we see that every bounded limit real number x has a unique *standard part* $\mathrm{st}(x)$ which differs from x by an *infinitesimal*, i.e., a limit real number of the form $\lim_{\alpha \to \alpha_\infty} x_\alpha$ where x_α converges to zero in the classical sense. As a consequence, the standard part $\mathrm{st}(d)$ of the limit metric function $d \colon X \times X \to {}^*\mathbf{R}$ is a genuine metric function $\mathrm{st}(d) \colon X \times X \to \mathbf{R}$. The resulting metric space $(X, \mathrm{st}(d))$ is often referred to as an *ultralimit* of the original metric spaces (X_α, d_α), although strictly speaking this conflicts slightly with the notation here, because we consider (X, d) to be the ultralimit instead.

2.1.2. Application: quantitative algebraic geometry. As a sample application of the above machinery, we shall use ultrafilter analysis to quickly deduce some quantitative (but not explicitly effective) algebraic geometry results from their more well-known qualitative counterparts. Significantly stronger results than the ones given here can be provided by the field of *effective algebraic geometry*, but that theory is somewhat more complicated than the classical qualitative theory, and the point to stress here is that one can obtain a "cheap" version of this effective algebraic geometry from the qualitative theory by a straightforward ultrafilter argument. There does not seem to be a comparably easy way to get such ineffective quantitative results without the use of ultrafilters or closely related tools (e.g., non-standard analysis or elementary limits).

We begin by recalling a basic definition:

Definition 2.1.6 (Algebraic set)**.** An (affine) algebraic set over an algebraically closed field k is a subset of k^n, where n is a positive integer, of the form

(2.1) $$\{x \in k^n : P_1(x) = \cdots = P_m(x) = 0\}$$

where $P_1, \ldots, P_m \colon k^n \to k$ are a finite collection of polynomials.

2.1. Ultralimit analysis

Now we turn to the quantitative theory, in which we try to control the *complexity* of various objects. Let us say that an algebraic set in k^n has complexity at most M if $n \leq M$, and one can express the set in the form (2.1) where $m \leq M$, and each of the polynomials $P_1, \ldots, P_m$ has degree at most M. We can then ask the question of to what extent one can make the above qualitative algebraic statements quantitative. For instance, it is known that a dimension 0 algebraic set is finite; but can we bound *how* finite it is in terms of the complexity M of that set? We are particularly interested in obtaining bounds here which are uniform in the underlying field k.

One way to do so is to open up an algebraic geometry textbook and carefully go through the *proofs* of all the relevant qualitative facts, and carefully track the dependence on the complexity. For instance, one could bound the cardinality of a dimension 0 algebraic set using *Bézout's theorem*. But here, we will use ultralimit analysis to obtain such quantitative analogues "for free" from their qualitative counterparts. The catch, though, is that the bounds we obtain are *ineffective*; they use the qualitative facts as a "black box", and one would have to go through the proof of these facts in order to extract anything better.

To begin the application of ultrafilter analysis, we use the following simple lemma.

Lemma 2.1.7 (Ultralimits of bounded complexity algebraic sets are algebraic). *Let n be a dimension. Suppose we have a sequence of algebraic sets $A_\alpha \subset k_\alpha^n$ over algebraically closed fields k_α, whose complexity is bounded by a quantity M which is uniform in α. Then if we set $k := \lim_{\alpha \to \alpha_\infty} k_\alpha$ and $A := \lim_{\alpha \to \alpha_\infty} A_\alpha$, then k is an algebraically closed field and $A \subset k^n$ is an algebraic set (also of complexity at most M).*

Conversely, every algebraic set in k^n is the ultralimit of algebraic sets in k_α^n of bounded complexity.

Proof. The fact that k is algebraically closed comes from Exercise 2.1.2. Now we look at the algebraic sets A_α. By adding dummy polynomials, if necessary, we can write

$$A_\alpha = \{x \in k_\alpha^n : P_{\alpha,1}(x) = \cdots = P_{\alpha,M}(x) = 0\}$$

where the $P_{\alpha,1}, \ldots, P_{\alpha,M} : k_\alpha^n \to k_\alpha$ of degree at most M.

We can then take ultralimits of the $P_{\alpha,i}$ to create polynomials $P_1, \ldots, P_M : k^n \to k$ of degree at most M. One easily verifies on taking ultralimits that

$$A = \{x \in k^n : P_1(x) = \cdots = P_M(x) = 0\},$$

and the first claim follows. The converse claim is proven similarly. □

Ultralimits preserve a number of key algebraic concepts (basically because such concepts are definable in first-order logic). We first illustrate this with the algebraic geometry concept of *dimension*. It is known that every non-empty algebraic set V in k^n has a *dimension* $\dim(V)$, which is an integer between 0 and n, with the convention that the empty set has dimension -1. There are many ways to define this dimension, but one way is to proceed by induction on the dimension n as follows. A non-empty algebraic subset of k^0 has dimension 0. Now if $n \geq 1$, we say that an algebraic set V has dimension d for some $0 \leq d \leq n$ if the following statements hold:

(i) For all but finitely many $t \in k$, the slice $V_t := \{x \in k^{n-1} : (x,t) \in V\}$ either all have dimension $d-1$, or are all empty.

(ii) For the remaining $t \in k$, the slice V_t has dimension at most d. If the generic slices V_t were all empty, then one of the exceptional V_t has to have dimension exactly d.

Informally, A has dimension d iff a generic slice of A has dimension $d-1$.

It is a non-trivial fact to show that every algebraic set in k^n does indeed have a well-defined dimension between -1 and n.

Now we see how dimension behaves under ultralimits.

Lemma 2.1.8 (Continuity of dimension). *Suppose that $A_\alpha \subset k_\alpha^n$ are algebraic sets over various algebraically closed fields k_α of uniformly bounded complexity, and let $A := \lim_{\alpha \to \alpha_\infty} A_\alpha$ be the limiting algebraic set given by Lemma 2.1.7. Then $\dim(A) = \lim_{\alpha \to \alpha_\infty} \dim(A_\alpha)$. In other words, we have $\dim(A) = \dim(A_\alpha)$ for all α sufficiently close to α_∞.*

Proof. One could obtain this directly from *Los's theorem*, but it is instructive to do this from first principles.

We induct on dimension n. The case $n = 0$ is trivial, so suppose that $n \geq 1$ and the claim has already been shown for $n - 1$. Write d for the dimension of A. If $d = -1$, then A is empty and so A_α must be empty for all α sufficiently close to α_∞, so suppose that $d \geq 0$. By the construction of dimension, the slice A_t all have dimension $d-1$ (or are all empty) for all but finitely many values $t_1, \ldots, t_r$ of $t \in k$. Let us assume that these generic slices A_t all have dimension $d-1$; the other case is treated similarly and is left to the reader. As k is the ultralimit of the k_α, we can write $t_i = \lim_{\alpha \to \alpha_\infty} t_{\alpha,i}$ for each $1 \leq i \leq r$. We claim that for α sufficiently close to α_∞, the slices $(A_\alpha)_{t_\alpha}$ have dimension $d-1$ whenever $t_\alpha \neq t_{\alpha,1}, \ldots, t_{\alpha,r}$. Indeed, suppose that this were not the case. Carefully negating the quantifiers (and using the ultrafilter property), we see that for α sufficiently close to α_∞, we can find $t_\alpha \neq t_{\alpha,1}, \ldots, t_{\alpha,r}$ such that $(A_\alpha)_{t_\alpha}$ has dimension different from $d-1$.

2.1. Ultralimit analysis

Taking ultralimits and writing $t := \lim_{\alpha \to \alpha_\infty} t_\alpha$, we see from the induction hypothesis that A_t has dimension different from $d - 1$, contradiction.

We have shown that for α sufficiently close to α_∞, all but finitely many slices of A_α have dimension $d - 1$, and thus by the definition of dimension, A_α has dimension d, and the claim follows. □

We can use this to deduce quantitative algebraic geometry results from qualitative analogues. For instance, from the definition of dimension we have

Lemma 2.1.9 (Qualitative Bezout-type theorem). *Every dimension 0 algebraic variety is finite.*

Using ultrafilter analysis, we immediately obtain the following quantitative analogue:

Lemma 2.1.10 (Quantitative Bezout-type theorem). *Let $A \subset k^n$ be an algebraic set of dimension 0 and complexity at most M over a field k. Then the cardinality A is bounded by a quantity C_M depending only on M (in particular, it is independent of k).*

Proof. By passing to the algebraic closure, we may assume that k is algebraically closed.

Suppose this were not the case. Carefully negating the quantifiers (and using the axiom of choice), we may find a sequence $A_\alpha \subset k_\alpha^n$ of dimension 0 algebraic sets and uniformly bounded complexity over algebraically closed fields k_α, such that $|A_\alpha| \to \infty$ as $\alpha \to \infty$. We pass to an ultralimit to obtain a limit algebraic set $A := \lim_{\alpha \to \alpha_\infty} A_\alpha$, which by Lemma 2.1.8 has dimension 0, and is thus finite by Lemma 2.1.9. But then this forces A_α to be bounded for α sufficiently close to α_∞ (indeed we have $|A_\alpha| = |A|$ in such a neighbourhood), contradiction. □

Remark 2.1.11. Note that this proof gives absolutely no bound on C_M in terms of M! One can get such a bound by using more effective tools, such as the actual Bezout theorem, but this requires more actual knowledge of how the qualitative algebraic results are proved. If one only knows the qualitative results as a black box, then the ineffective quantitative result is the best one can do.

Now we give another illustration of the method. The following fundamental result in algebraic geometry is known:

Lemma 2.1.12 (Qualitative Noetherian condition). *There does not exist an infinite decreasing sequence of algebraic sets in an affine space k^n, in which each set is a proper subset of the previous one.*

Using ultralimit analysis, one can convert this qualitative result into an ostensibly stronger quantitative version:

Lemma 2.1.13 (Quantitative Noetherian condition). *Let $F\colon \mathbf{N} \to \mathbf{N}$ be a function. Let $A_1 \supsetneq A_2 \supsetneq \cdots \supsetneq A_R$ be a sequence of properly nested algebraic sets in k^n for some algebraically closed field k, such that each A_i has complexity at most $F(i)$. Then R is bounded by C_F for some C_F depending only on F (in particular, it is independent of k).*

Remark 2.1.14. Specialising to the case when F is a constant M, we see that there is an upper bound on properly nested sequences of algebraic sets of bounded complexity; but the statement is more powerful than this because we allow F to be non-constant. Note that one can easily use this strong form of the quantitative Noetherian condition to recover Lemma 2.1.12 (why?), but if one only knew Lemma 2.1.13 in the constant case $F = M$, then this does not obviously recover Lemma 2.1.12.

Proof. Note that n is bounded by $F(1)$, so it will suffice to prove this claim for a fixed n.

Fix n. Suppose the claim failed. Carefully negating all the quantifiers (and using the axiom of choice), we see that there exists an F, a sequence k_α of algebraically closed fields, a sequence R_α going to infinity, and sequences

$$A_{\alpha,1} \supsetneq \cdots \supsetneq A_{\alpha,R_\alpha}$$

of properly nested algebraic sets in k_α^n, with each $A_{\alpha,i}$ having complexity at most $F(i)$.

We take an ultralimit of everything that depends on α, creating an algebraically closed field $k = \lim_{\alpha \to \alpha_\infty} k_\alpha$, and an infinite sequence[6]

$$A_1 \supsetneq A_2 \supsetneq \cdots$$

of properly nested algebraic sets in k^n. But this contradicts Lemma 2.1.12. □

Again, this argument gives absolutely no clue as to how C_F is going to depend on F.

Let us give one last illustration of the ultralimit analysis method, which contains an additional subtlety. Define an *algebraic variety* to be an algebraic set which is *irreducible*, which means that it cannot be expressed as the union of two proper subalgebraic sets. This notation is stable under ultralimits:

---

[6]In fact, we could continue this sequence into a limit sequence up to the unbounded limit number $\lim_{\alpha \to \alpha_\infty} R_\alpha$, but we will not need this *overspill* here.

2.1. Ultralimit analysis

Lemma 2.1.15 (Continuity of irreducibility). *Suppose that $A_\alpha \subset k_\alpha^n$ are algebraic sets over various algebraically closed fields k_α of uniformly bounded complexity, and let $A := \lim_{\alpha \to \alpha_\infty} A_\alpha$ be the limiting algebraic set given by Lemma 2.1.7. Then A is an algebraic variety if and only if A_α is an algebraic variety for all α sufficiently close to α_∞.*

However, this lemma is somewhat harder to prove than previous ones, because the notion of irreducibility is not quite a first-order statement. The following exercises show the limit of what one can do without using some serious algebraic geometry:

Exercise 2.1.3. Let the notation and assumptions be as in Lemma 2.1.15. Show that if A is *not* an algebraic variety, then A_α is not an algebraic variety for all α sufficiently close to α_∞.

Exercise 2.1.4. Let the notation and assumptions be as in Lemma 2.1.15. Call an algebraic set *M-irreducible* if it cannot be expressed as the union of two proper algebraic sets of complexity at most M. Show that if A is an algebraic variety, then for every $M \geq 1$, A_α is M-irreducible for all α sufficiently close to α_∞.

These exercises are not quite strong enough to give Lemma 2.1.15, because M-irreducibility is a weaker concept than irreducibility. However, one can do better by applying some further facts in algebraic geometry. Given an algebraic set A of dimension $d \geq 0$ in an affine space k^n, one can assign a *degree* $\deg(A)$, which is a positive integer such that $|A \cap V| = \deg(A)$ for *generic* $n - d$-dimensional affine subspaces of k^n, which means that V belongs to the *affine Grassmannian* Gr of $n-d$-dimensional affine subspaces of k^n, after removing an algebraic subset of Gr of dimension strictly less than that of Gr. It is a standard fact of algebraic geometry that every algebraic set can be assigned a degree. Somewhat less trivially, the degree controls the complexity:

Theorem 2.1.16 (Degree controls complexity). *Let A be an algebraic variety of k^n of degree D. Then A has complexity at most $C_{n,D}$ for some constants n, D depending only on n, D.*

Proof. It[7] suffices to show that A can be cut out by polynomials of degree D, since the space of polynomials of degree D that vanish on A is a vector space of dimension bounded only by n and D.

Let A have dimension d. We pick a generic affine subspace V of k^n of dimension $n - d - 2$, and consider the cone $C(V, A)$ formed by taking the union of all the lines joining a point in V to a point in A. This is an algebraic

---
[7] We thank Jordan Ellenberg and Ania Otwinowska for this argument, which goes back to [Mu1970].

image of $V \times A \times \mathbf{R}$ and is thus generically an algebraic set of dimension $n-1$, i.e., a hypersurface. Furthermore, as A has degree D, it is not hard to see that $C(V, A)$ has degree D as well. Since a hypersurface is necessarily cut out by a single polynomial, this polynomial must have degree D.

To finish the claim, it suffices to show that the intersection of the $C(V, A)$ as V varies is exactly A. Clearly, this intersection contains A. Now let p be any point not in A. The cone of A over p can be viewed as an algebraic subset of the projective space P^{n-1} of dimension d; meanwhile, the cone of a generic subspace V of dimension $n-d-2$ is a generic subspace of P^{n-1} of the same dimension. Thus, for generic V, these two cones do not intersect, and thus p lies outside of $C(V, A)$, and the claim follows. □

Remark 2.1.17. There is a stronger theorem that asserts that if the degree of a *scheme* in k^n is bounded, then the complexity of that scheme is bounded as well. The main difference between a variety and a scheme here is that for a scheme, we not only specify the set of points cut out by the scheme, but also the ideal of functions that we want to think of as vanishing on that set. This theorem is significantly more difficult than the above result; see [**Kl1971**, Corollary 6.11].

Given this theorem, we can now prove Lemma 2.1.15.

Proof. In view of Exercise 2.1.3, it suffices to show that if A is irreducible, then the A_α are irreducible for α sufficiently close to α_0.

The algebraic set A has some dimension d and degree D, thus $|A \cap V| = D$ for generic affine $n-d$-dimensional subspaces V of k^n. Undoing the limit using Lemma 2.1.7 and Lemma 2.1.8 (adapted to the Grassmannian Gr rather than to affine space), we see that for α sufficiently close to α_0, $|A_\alpha \cap V_\alpha| = D$ for generic affine $n-d$-dimensional subspaces V_α of k_α^n. In other words, A_α has degree D, and thus by Theorem 2.1.16, any algebraic variety of A_α of the same dimension d as A_α will have complexity bounded by $C_{n,D}$ uniformly in α. Let B_α be a d-dimensional algebraic subvariety of A_α, and let B be the ultralimit of the B_α. Then by Lemma 2.1.7, Lemma 2.1.8 and the uniform complexity bound, B is a d-dimensional algebraic subset of A, and thus must equal all of A by irreducibility of A. But this implies that $B_\alpha = A_\alpha$ for all α sufficiently close to α_0, and the claim follows. □

We give a sample application of this result. From the Noetherian condition we easily obtain

Lemma 2.1.18 (Qualitative decomposition into varieties). *Every algebraic set can be expressed as a union of finitely many algebraic varieties.*

Using ultralimit analysis, we can make this quantitative:

2.1. Ultralimit analysis

Lemma 2.1.19 (Quantitative decomposition into varieties). *Let $A \subset k^n$ be an algebraic set of complexity at most M over an algebraically closed field k. Then A can be expressed as the union of at most C_M algebraic varieties of complexity at most C_M, where C_M depends only on M.*

Proof. As n is bounded by M, it suffices to prove the claim for a fixed n.

Fix n and M. Suppose the claim failed. Carefully negating all the quantifiers (and using the axiom of choice), we see that there exists a sequence $A_\alpha \subset k_\alpha^n$ of uniformly bounded complexity, such that A_α cannot be expressed as the union of at most α algebraic varieties of complexity at most α. Now we pass to an ultralimit, obtaining a limit algebraic set $A \subset k^n$. As discussed earlier, A is an algebraic set over an algebraically closed field and is thus expressible as the union of a finite number of algebraic varieties $A_1, \ldots, A_m$. By Lemma 2.1.7 and Lemma 2.1.15, each A_i is an ultralimit of algebraic varieties $A_{\alpha,i}$ of bounded complexity. The claim follows. $\square$

2.1.3. Application: Quantitative Gromov theorem. As a further illustration of ultralimit analysis, we now establish the correspondence principle between finitary and infinitary forms of the following famous theorem of Gromov [**Gr1981**]:

Theorem 2.1.20 (Qualitative Gromov theorem). *Every finitely generated group of polynomial growth is virtually nilpotent.*

Let us now make the observation (already observed in [**Gr1981**]) that this theorem implies (and is in fact equivalent to) a quantitative version:

Theorem 2.1.21 (Quantitative Gromov theorem). *For every C, d there exists R such that if G is generated by a finite set S with the growth condition $|B_S(r)| \leq Cr^d$ for all $1 \leq r \leq R$, then G is virtually nilpotent, and furthermore it has a nilpotent subgroup of step and index at most $M_{C,d}$ for some $M_{C,d}$ depending only on C, d. Here $B_S(r)$ is the ball of radius r generated by the set S.*

Proof. We use ultralimit analysis. Suppose this theorem failed. Carefully negating the quantifiers, we find that there exists C, d, as well as a sequence G_α of groups generated by a finite set S_α such that $|B_{S_\alpha}(r)| \leq Cr^d$ for all $1 \leq r \leq \alpha$, and such that G_α does not contain any nilpotent subgroup of step and index at most α.

Now we take ultralimits, setting $G := \lim_{\alpha \to \alpha_\infty} G_\alpha$ and $S := \lim_{\alpha \to \alpha_\infty} S_\alpha$. As the S_α have cardinality uniformly bounded (by Cr^1), S is finite. The set S need not generate G, but it certainly generates some subgroup $\langle S \rangle$ of this group. Since $|B_{S_\alpha}(r)| \leq Cr^d$ for all α and all $1 \leq r \leq \alpha$, we see on taking ultralimits that $|B_S(r)| \leq Cr^d$ for all r. Thus $\langle S \rangle$ is of polynomial growth, and is thus virtually nilpotent.

Now we need to undo the ultralimit, but this requires a certain amount of preparation. We know that $\langle S \rangle$ contains a finite index nilpotent subgroup G'. As $\langle S \rangle$ is finitely generated, the finite index subgroup G' is also[8]. Let S' be a set of generators for G'. Since G' is nilpotent of some step s, all commutators of S' of length at least $s+1$ vanish.

Writing S' as an ultralimit of S'_α, we see that the S'_α are finite subsets of G_α which generate some subgroup G'_α. Since all commutators of S' of length at least $s+1$ vanish, the same is true for S'_α for α close enough to α_∞, and so G'_α is nilpotent for such α with step bounded uniformly in α.

Finally, if we let R be large enough that $B_S(R)$ intersects every coset of G', then we can cover $B_S(R+1)$ by a product of $B_S(R)$ and some elements of G' (which are of course finite products of elements in S' and their inverses). Undoing the ultralimit, we see that for α sufficiently close to α_∞, we can cover $B_{S_\alpha}(R+1)$ by the product of $B_{S_\alpha}(R)$ and some elements of G'_α. Iterating this we see that we can cover all of G_α by $B_{S_\alpha}(R)$ times G'_α, and so G'_α has finite index bounded uniformly in α. But this contradicts the construction of G_α. $\square$

Remark 2.1.22. As usual, the argument gives no effective bound on $M_{C,d}$. Obtaining such an effective bound is in fact rather non-trivial; see [**Sh2009**] for further discussion.

2.1.4. Application: Furstenberg correspondence principle.

Let me now redo another application of the correspondence principle via ultralimit analysis. We will begin with the following famous result of Furstenberg [**Fu1977**]:

Theorem 2.1.23 (Furstenberg recurrence theorem). *Let $(X, \mathcal{B}, \mu, T)$ be a measure-preserving system, and let $A \subset X$ have positive measure. Let $k \geq 1$. Then there exists $r > 0$ such that $A \cap T^r A \cap \cdots \cap T^{(k-1)r} A$ is non-empty.*

We then use this theorem and ultralimit analysis to derive the following well-known result of Szemerédi [**Sz1975**]:

Theorem 2.1.24 (Szemerédi's theorem). *Every set of integers of positive upper density contains arbitrarily long arithmetic progressions.*

---

[8]Here is a quick proof of this claim: for R large enough, $B_S(R)$ will intersect every coset of G'. As a consequence, one can describe the action of $\langle S \rangle$ on the finite set $\langle S \rangle / G'$ using only knowledge of $B_S(2R+1) \cap G'$. In particular, $B_S(2R+1) \cap G'$ generates a finite index subgroup. Increasing R, the index of this subgroup is non-increasing, and thus must eventually stabilise. At that point, we generate all of G'.

2.1. Ultralimit analysis

Proof. Suppose this were not the case. Then there exists $k \geq 1$ and a set A of positive upper density with no progressions of length k. Unpacking the definition of positive upper density, this means that there exists $\delta > 0$ and a sequence $N_\alpha \to \infty$ such that

$$|A \cap [-N_\alpha, N_\alpha]| \geq \delta |[-N_\alpha, N_\alpha]|$$

for all α. We pass to an ultralimit, introducing the limit natural number $N := \lim_{\alpha \to \alpha_\infty} N_\alpha$ and using the ultrapower $^*A = \lim_{\alpha \to \alpha_\infty} A$ (note that A is a space, not an ordinary object). Then we have

$$|^*A \cap [-N, N]| \geq \delta |[-N, N]|$$

where the cardinalities are in the limit sense. Note also that $^*A$ has no progressions of length k.

Consider the space of all Boolean combinations of shifts $^*A + r$ of $^*A$, where r ranges over (standard) integers; thus, for instance,

$$(^*A + 3) \cap (^*A + 5) \setminus (^*A - 7)$$

would be such a set. We call such sets *definable sets*. We give each such definable set B a limit measure

$$\mu(B) := |B \cap [-N, N]|/[-N, N].$$

This measure takes values in the limit interval $^*[0, 1]$ and is clearly a finitely additive probability measure. It is also nearly translation invariant in the sense that

$$\mu(B + k) = \mu(B) + o(1)$$

for any standard integer k, where $o(1)$ is an *infinitesimal* (i.e., a limit real number which is smaller in magnitude than any positive standard real number). In particular, the standard part $\text{st}(\mu)$ of μ is a finitely additive *standard* probability measure. Note from construction that $\text{st}(\mu)(A) \geq \delta$.

Now we convert this finitely additive measure into a countably additive one. Let $2^{\mathbf{Z}}$ be the set of all subsets B of the integers. This is a compact metrisable space, which we endow with the Borel σ-algebra $\mathcal{B}$ and the standard shift $T \colon B \mapsto B + 1$. The Borel σ-algebra is generated by the clopen sets in this space, which are Boolean combinations of $T^r E$, where E is the basic *cylinder set* $E := \{B \in 2^{\mathbf{Z}} : 0 \in B\}$. Each clopen set can be assigned a definable set in $^*\mathbf{Z}$ by mapping $T^r E$ to $^*A + r$ and then extending by Boolean combinations. The finitely additive probability measure $\text{st}(\mu)$ on definable sets then pulls back to a finitely additive probability measure ν on clopen sets in $2^{\mathbf{Z}}$. Applying the *Carathéodory extension theorem* (see e.g. [**Ta2011**, §1.7]), taking advantage of the compactness of $2^{\mathbf{Z}}$, we can extend this finitely additive measure to a countably additive Borel probability measure.

By construction, $\nu(E) \geq \delta > 0$. Applying Theorem 2.1.23, we can find $r > 0$ such that $E \cap T^r E \cap \cdots \cap T^{(k-1)r} E$ is non-empty. This implies that $^*A \cap (^*A + r) \cap \cdots \cap (^*A + (k-1)r)$ is non-empty, and so $^*A$ contains an arithmetic progression of length k, a contradiction. □

Remark 2.1.25. The above argument is nearly identical to the usual proof of the correspondence principle, which uses *Prokhorov's theorem* (see e.g. [**Ta2010**, §1.10]) instead of ultrafilters. The measure constructed above is essentially the *Loeb measure* [**Lo1975**] for the ultraproduct.

2.1.5. Relationship with non-standard analysis. Ultralimit analysis is extremely close to, but subtly different from, non-standard analysis, because of a shift of emphasis and philosophy. The relationship can be illustrated by the following table of analogies:

Digits	Strings of digits	Numbers
Symbols	Strings of symbols	Sentences
Set theory	Finite von Neumann ordinals	Peano arithmetic
Rational numbers **Q**	$\overline{\mathbf{Q}}$	Real numbers **R**
Real analysis	Analysis on $\overline{\mathbf{R}}$	Complex analysis
R	$\mathbf{R}^2$	Euclidean plane geometry
R	Coordinate chart atlases	Manifolds
R	Matrices	Linear transformations
Algebra	Sheaves of rings	Schemes
Deterministic theory	Measure theory	Probability theory
Probability theory	Von Neumann algebras	Noncommutative prob. theory
Classical mechanics	Hilbert space mechanics	Quantum mechanics
Finitary analysis	Asymptotic analysis	Infinitary analysis
Combinatorics	Correspondence principle	Ergodic theory
Quantitative analysis	Compactness arguments	Qualitative analysis
Standard analysis	Ultralimit analysis	Non-standard analysis

Here $\overline{\mathbf{R}}$ is the algebraic completion of the reals, but $\overline{\mathbf{Q}}$ is the metric completion of the rationals.

In the first column one has a "base" theory or concept, which implicitly carries with it a certain ontology and way of thinking, regarding what objects one really cares to study, and what objects really "exist" in some mathematical sense. In the second column one has a fancier theory than the base theory (typically a "limiting case", a "generalisation", or a "completion" of the base theory), but one which still shares a close relationship with the base theory, in particular, largely retaining the ontological and conceptual mindset of that theory. In the third column one has a new theory, which is *modeled* by the theories in the middle column, but which is not tied to that model, or to the implicit ontology and viewpoint carried by that model. For instance, one *can* think of a complex number as an element of the algebraic completion of the reals, but one does not *have* to, and indeed in many parts of complex analysis or complex geometry one wants to ignore the role of the reals as much as possible. Similarly for other rows of the above table. See, for instance, [**Ta2011b**, §1.1] for further discussion of the distinction between measure theory and probability theory.

Remark 2.1.26. The relationship between the second and third columns of the above table is also known as the *map-territory relation*.

Returning to ultralimit analysis, this is a type of analysis which still shares close ties with its base theory, standard analysis, in that all the objects one considers are either standard objects, or ultralimits of such objects (and similarly for all the spaces one considers). But more importantly, one continues to *think of* non-standard objects as being ultralimits of standard objects, rather than having an existence which is largely independent of the concept of base theory of standard analysis. This perspective is reversed in non-standard analysis: one views the non-standard universe as existing in its own right, and the fact that the standard universe can be embedded inside it is a secondary feature (albeit one which is absolutely essential if one is to use non-standard analysis in any non-trivial manner to say something new about standard analysis). In non-standard analysis, ultrafilters are viewed as one tool in which one can construct the non-standard universe from the standard one, but their role in the subject is otherwise minimised. In contrast, the ultrafilter α_∞ plays a prominent role in ultralimit analysis.

In my opinion, none of the three columns here are inherently "better" than the other two; but they do work together quite well. In particular, the middle column serves as a very useful bridge to carry results back and forth between the worlds of the left and right columns.

2.2. Higher order Hilbert spaces

Recall that a (complex, semi-definite) *inner product space* is a complex vector space V equipped with a sesquilinear form $\langle,\rangle \colon V \times V \to \mathbf{C}$ which is conjugate symmetric, in the sense that $\langle w, v \rangle = \overline{\langle v, w \rangle}$ for all $v, w \in V$, and non-negative in the sense that $\langle v, v \rangle \geq 0$ for all $v \in V$. By inspecting the non-negativity of $\langle v + \lambda w, v + \lambda w \rangle$ for complex numbers $\lambda \in \mathbf{C}$, one obtains the *Cauchy-Schwarz inequality*

$$|\langle v, w \rangle| \leq |\langle v, v \rangle|^{1/2} |\langle w, w \rangle|^{1/2};$$

if one then defines $\|v\| := |\langle v, v \rangle|^{1/2}$, one then quickly concludes the *triangle inequality*

$$\|v + w\| \leq \|v\| + \|w\|$$

which then soon implies that $\|\|$ is a *semi-norm*[9] on V. If we make the additional assumption that the inner product $\langle,\rangle$ is positive definite, i.e., that $\langle v,v\rangle > 0$ whenever v is non-zero, then this semi-norm becomes a norm. If V is complete with respect to the metric $d(v,w) := \|v - w\|$ induced by this norm, then V is called a *Hilbert space*.

The above material is extremely standard, and can be found in any graduate real analysis text (e.g. [**Ta2010**, §1.6]). But what is perhaps less well known (except inside the fields of additive combinatorics and ergodic theory) is that the above theory of classical Hilbert spaces is just the first case of a hierarchy of *higher order Hilbert spaces*, in which the binary inner product $f, g \mapsto \langle f, g\rangle$ is replaced with a 2^d-ary inner product $(f_\omega)_{\omega \in \{0,1\}^d} \mapsto \langle(f_\omega)_{\omega \in \{0,1\}^d}\rangle$ that obeys an appropriate generalisation of the conjugate symmetry, sesquilinearity, and positive semi-definiteness axioms. Such inner products then obey a higher order Cauchy-Schwarz inequality, known as the *Cauchy-Schwarz-Gowers* inequality, and then also obey a triangle inequality and become semi-norms (or norms, if the inner product was non-degenerate). Examples of such norms and spaces include the *Gowers uniformity norms* $\|\|\|_{U^d(G)}$, the *Gowers box norms* $\|\|\|_{\Box^d(X_1 \times \cdots \times X_d)}$, and the *Gowers-Host-Kra semi-norms* $\|\|\|_{U^d(X)}$; a more elementary example are the family of Lebesgue spaces $L^{2^d}(X)$ when the exponent is a power of two. They play a central role in modern additive combinatorics and to certain aspects of ergodic theory, particularly those relating to Szemerédi's theorem (or its ergodic counterpart, the Furstenberg multiple recurrence theorem); they also arise in the regularity theory of hypergraphs (which is not unrelated to the other two topics).

A simple example to keep in mind here is the order two Hilbert space $L^4(X)$ on a measure space $X = (X, \mathcal{B}, \mu)$, where the inner product takes the form

$$\langle f_{00}, f_{01}, f_{10}, f_{11}\rangle_{L^4(X)} := \int_X f_{00}(x)\overline{f_{01}(x)f_{10}(x)}f_{11}(x)\ d\mu(x).$$

In this section we will set out the abstract theory of such higher order Hilbert spaces; this is drawn from the more concrete work of Gowers [**Go2001**] and Host-Kra [**HoKr2005**], but this material is actually quite abstract, and is not particularly tied to any explicit choice of norm so long as a certain axioms are satisfied. In applications, one can (and probably

---

[9]A *semi-norm* on a vector space V is a map $v \mapsto \|v\|$ from V to the non-negative reals $[0, +\infty)$ which obeys the triangle inequality $\|v + w\| \le \|v\| + \|w\|$ and the homogeneity relation $\|cv\| = |c|\|v\|$ for all $v, w \in V$ and $c \in \mathbf{C}$. A *norm* is a semi-norm with the additional property that $\|v\| > 0$ for all non-zero v.

2.2. Higher order Hilbert spaces

should) work in the concrete setting, but we will record the abstract axiomatic approach here, as this does not appear to be explicitly in the literature elsewhere.

2.2.1. Definition of a higher order Hilbert space. Let V, W be complex vector spaces. Then one can form the (algebraic) *tensor product* $V \otimes W$, which can be defined as the vector space spanned by formal tensor products $v \otimes w$, subject to the constraint[10] that the tensor product is bilinear (i.e. that $v \otimes (w_1 + w_2) = (v \otimes w_1) + (v \otimes w_2)$, $v \otimes cw = c(v \otimes w)$, and similarly with the roles of v and w reversed). More generally, one can define the tensor product $\bigotimes_{\omega \in \Omega} V_\omega$ of any finite family of complex vector spaces V_ω.

Given a complex vector space V, one can define its *complex conjugate*[11] $\overline{V}$ to be the set of formal conjugates $\{\overline{v} : v \in V\}$ of vectors in V, with the vector space operations given by

$$0 := \overline{0},$$
$$\overline{v} + \overline{w} := \overline{v + w},$$
$$c\overline{v} := \overline{\overline{c}v}.$$

The map $v \mapsto \overline{v}$ is then an antilinear isomorphism from V to $\overline{V}$. We adopt the convention that $\overline{\overline{v}} = v$, thus $v \mapsto \overline{v}$ is also an antilinear isomorphism from $\overline{V}$ to V.

For inductive reasons, it is convenient to use finite sets A of labels, rather than natural numbers d, to index the order of the systems we will be studying. In any case, the cardinality $|A|$ of the set of labels will be the most important feature of this set.

Given a complex vector space V and a finite set A of labels, we form the tensor cube $V^{[A]}$ to be

$$V^{[A]} := \bigotimes_{\omega \in \{0,1\}^A} \mathcal{C}^{|\omega|} V,$$

where $\mathcal{C}$ is the conjugation map $V \mapsto \overline{V}$, and $|\omega| := \sum_{i \in A} \omega_i$ when $\omega = (\omega_i)_{i \in A}$; thus, for instance[12], $V^{[\{\}]} = V$, $V^{[\{1\}]} \equiv V \otimes \overline{V}$ is spanned by tensor products $v_0 \otimes \overline{v_1}$ with $v_0, v_1 \in V$, $V^{[\{1,2\}]} \equiv V \otimes \overline{V} \otimes \overline{V} \otimes V$ is spanned by tensor products $v_{00} \otimes \overline{v_{01}} \otimes \overline{v_{10}} \otimes v_{11}$ with $v_{00}, v_{01}, v_{10}, v_{11} \in V$, and so forth.

---

[10] More formally, one would quotient out by the subspace generated by elements such as $v \otimes (w_1 + w_2) - (v \otimes w_1) - (v \otimes w_2)$ or $v \otimes cw - c(v \otimes w)$ to create the tensor product.

[11] One can work with real higher order Hilbert spaces instead of complex ones, in which case the conjugation symbols can be completely ignored.

[12] It would be better to order the four factors $v_{00}, v_{01}, v_{10}, v_{11}$ in a square pattern, rather than linearly as is done here, but we have used the inferior linear ordering here for typographical reasons.

Given any finite set A of labels and any $i \in A$, one can form an identification
$$V^{[A]} \equiv V^{[A\setminus\{i\}]} \otimes \overline{V^{[A\setminus\{i\}]}}$$
by identifying a tensor product $\bigotimes_{\omega \in \{0,1\}^A} C^{|\omega|} v_\omega$ in $V^{[A]}$ with

$$\left(\bigotimes_{\omega' \in \{0,1\}^{A\setminus i}} C^{|\omega'|} v_{(\omega',0)} \right) \otimes \overline{\left(\bigotimes_{\omega' \in \{0,1\}^{A\setminus i}} C^{|\omega'|} v_{(\omega',1)} \right)}$$

where, for $\omega' \in \{0,1\}^{A\setminus i}$ and $\omega_i \in \{0,1\}$, (ω',ω_i) denotes the element of $\{0,1\}^A$ that agrees with ω' on $A\setminus i$ and equals ω_i on i. We refer to this identification as $\bigotimes_i$, thus

$$\bigotimes_i \colon V^{[A\setminus\{i\}]} \otimes \overline{V^{[A\setminus\{i\}]}} \to V^{[A]}$$

is an isomorphism, and one can define the ith tensor product $v \otimes_i \overline{w} \in V^{[A]}$ of two elements $v, w \in V^{[A\setminus\{i\}]}$. Thus, for instance, if $v = v_0 \otimes \overline{v_1}$ and $w = w_0 \otimes \overline{w_1}$ are elements of $V^{[\{1\}]}$, then

$$v \otimes_2 \overline{w} = v_0 \otimes \overline{v_1} \otimes \overline{w_0} \otimes w_1$$

using the linear ordering conventions used earlier. If, instead, we view v, w as elements of $V^{[\{2\}]}$ rather than $V^{[\{1\}]}$, then

$$v \otimes_1 \overline{w} = v_0 \otimes \overline{w_0} \otimes \overline{v_1} \otimes w_1.$$

A (semi-)definite inner product $\langle,\rangle$ on a complex vector space V can be viewed as a linear functional $\langle\rangle \colon V \otimes \overline{V} \to \mathbf{C}$ on $V^{[\{1\}]} = V \otimes \overline{V}$ obeying a conjugation symmetry and positive (semi-)definiteness property, defined on tensor products $v \otimes \overline{w}$ as $\langle v \otimes \overline{w} \rangle := \langle v, w \rangle$. With this notation, the conjugation symmetry axiom becomes

$$\langle w \otimes \overline{v} \rangle := \overline{\langle v \otimes \overline{w} \rangle}$$

and the positive semi-definiteness property becomes

$$\langle v \otimes \overline{v} \rangle \geq 0$$

with equality iff $v = 0$ in the definite case.

Now we can define a higher order inner product space.

Definition 2.2.1 (Higher order inner product space). Let A be a finite set of labels. A *(semi-definite) inner product space of order A* is a complex vector space V, together with a linear functional $\langle\rangle_A \colon V^{[A]} \to \mathbf{C}$ that obeys the following axiom:

- (Splitting axiom) For every $i \in A$, $\langle\rangle_A$ is a semi-definite classical inner product $\langle\rangle_{A\setminus\{i\}}$ on $V^{[A\setminus\{i\}]} \otimes \overline{V^{[A\setminus\{i\}]}}$, which we identify with $V^{[A]}$ using $\bigotimes_i$ as mentioned above.

2.2. Higher order Hilbert spaces

We say that the inner product space is *positive definite* if one has[13]

$$\langle \bigotimes_{\omega \in \{0,1\}^A} C^{|\omega|} v \rangle_A > 0$$

whenever $v \in V$ is non-zero.

For instance, if A is the empty set, then an inner product space of order A is just a complex vector space V equipped with a linear functional $v \mapsto \langle v \rangle_A$ from V to $\mathbf{C}$ (which one could interpret as an expectation or a trace, if one wished). If A is a singleton set, then an inner product space of order A is the same thing as a classical inner product space.

If $A = \{1, 2\}$, then an inner product space of order A is a complex vector space V equipped with a linear functional $\langle \rangle_A \colon V \otimes \overline{V} \otimes \overline{V} \otimes V$ which, in particular, gives rise to a quartisesquilinear (!) form

$$(v_{00}, v_{01}, v_{10}, v_{11}) \mapsto \langle v_{00} \otimes \overline{v_{01}} \otimes \overline{v_{10}} \otimes v_{11} \rangle_A$$

which is a classical inner product in two different ways; thus, for instance, we have

$$\langle v_{00} \otimes \overline{v_{01}} \otimes \overline{v_{10}} \otimes v_{11} \rangle_A = \langle v_{00} \otimes \overline{v_{01}}, v_{10} \otimes \overline{v_{11}} \rangle_{\{2\}}$$

for $v_{00}, v_{01}, v_{10}, v_{11} \in V$ and some classical inner product $\langle , \rangle_{\{2\}}$ on $V^{[\{2\}]}$, and similarly

$$\langle v_{00} \otimes \overline{v_{01}} \otimes \overline{v_{10}} \otimes v_{11} \rangle_A = \langle v_{00} \otimes \overline{v_{10}}, v_{01} \otimes \overline{v_{11}} \rangle_{\{1\}}$$

for some classical inner product $\langle , \rangle_{\{1\}}$ on $V^{[\{1\}]}$.

2.2.2. Examples. Let us now give the three major (and inter-related) examples of inner product spaces of higher order: the *Gowers uniformity spaces*, that arise in additive combinatorics; the *Gowers box spaces*, which arise in hypergraph regularity theory, and the *Gowers-Host-Kra spaces*, which arise in ergodic theory. We also remark on the much simpler example of the Lebesgue spaces of dyadic exponent.

The first example is the family of *Gowers uniformity spaces* $U^A(G)$, which we will define for simplicity on a finite additive group G (one can also define this norm more generally on finite subsets of abelian groups, and probably also nilpotent groups, but we will not do so here). Here A is a finite set of labels; in applications one usually sets $A := \{1, \ldots, d\}$, in which case one abbreviates $U^{\{1,\ldots,d\}}(G)$ as $U^d(G)$. The space $U^A(G)$ is the space of all functions $f \colon G \to \mathbf{C}$, and so $U^A(G)^{[A]}$ can be canonically identified

---

[13]Note from the splitting axiom that one already has the non-strict inequality. But the positive definiteness property is weaker than the assertion that each of the classical inner products are themselves non-degenerate.

with the space of functions $F\colon G^{\{0,1\}^A} \to \mathbf{C}$. To make $U^A(G)$ into an inner product space of order A, we define

$$\langle F \rangle_A := \mathbf{E}_{x \in G^{[A]}} F(x)$$

where $G^{[A]}$ is the subgroup of $G^{\{0,1\}^A}$ consisting of the parallelopipeds

$$G^{[A]} := \{(x + \sum_{i \in A} \omega_i h_i)_{\omega \in \{0,1\}^A} : x \in G, h_i \in G \text{ for all } i \in A\}.$$

This is clearly a linear functional. To verify the splitting axiom, one observes the identity

$$\langle F_0 \otimes_i \overline{F_1} \rangle_A = \mathbf{E}_{h_j \in G \text{ for } j \in A \setminus \{i\}} \mathbf{E}_{x, h_i \in G}$$

$$F_0((x + \sum_{j \in A \setminus \{i\}} \omega_j h_j)_{\omega \in \{0,1\}^{A \setminus \{i\}}})$$

$$\overline{F_1}((x + h_i + \sum_{j \in A \setminus \{i\}} \omega_j h_j)_{\omega \in \{0,1\}^{A \setminus \{i\}}})$$

for any $i \in A$ and $F_0, F_1 \in U^A(G)^{[A \setminus \{i\}]}$. The right-hand side is then a semi-definite classical inner product on $U^A(G)^{[A \setminus \{i\}]}$; the semi-definiteness becomes more apparent if one makes the substitution $(x, y) := (x, x + h_i)$.

Specialising to tensor products, we obtain the *Gowers inner product*

$$\langle \bigotimes_{\omega \in \{0,1\}^A} \mathcal{C}^{|\omega|} f_\omega \rangle_A = \mathbf{E}_{x \in G, h_i \in G \forall i \in A} \prod_{\omega \in \{0,1\}^A} \mathcal{C}^{|\omega|} f_\omega(x + \sum_{i=1}^A \omega_i h_i).$$

Thus, for instance, when $A = \{1, 2\}$,

$$\langle f_{00} \otimes \overline{f_{01}} \otimes \overline{f_{10}} \otimes f_{11} \rangle_A$$
$$= \mathbf{E}_{x, h_1, h_2 \in G} f_{00}(x) \overline{f_{10}}(x + h_2) \overline{f_{10}}(x + h_1) f_{11}(x + h_1 + h_2).$$

The second example is the family of the (incomplete) *Gowers box spaces* $\square^A \cap L^\infty(X)$, defined on a Cartesian product $X := \prod_{i \in A} X_i$ of a family $X_i = (X_i, \mathcal{B}_i, \mu_i)$ of measure spaces indexed by a finite set A. To avoid some minor technicalities regarding absolute integrability, we assume that all the measure spaces have finite measure (the theory also works in the σ-finite case, but we will not discuss this here). This space is the space of all bounded measurable functions $f \in L^\infty(X)$ (here, for technical reasons, it is best not to quotient out by almost everywhere equivalence until later in the theory). The tensor power $L^\infty(X)^{[A]}$ can thus be identified with a subspace of $L^\infty(X^{\{0,1\}^A})$ (roughly speaking, this is the subspace of "elementary functions"). We can then define an inner product of order A by the formula

$$\langle F \rangle = \int_X \int_X F(((x_{\omega_i, i})_{i \in A})_{\omega \in \{0,1\}^A}) \, d\mu(x_0) d\mu(x_1)$$

2.2. Higher order Hilbert spaces

for all $F \in L^\infty(X)^{[A]} \subset L^\infty(X^{\{0,1\}^A})$, where $x_0 = (x_{0,i})_{i \in A}$ and $x_1 = (x_{0,i})_{i \in A}$ are integrated using product measure $\mu := \prod_{i \in A} \mu_i$.

The verification of the splitting property is analogous to that for the Gowers uniformity spaces. Indeed, there is the identity

$$\langle F_0 \otimes_i \overline{F_1} \rangle_A = \int_{X^{(i)}} \int_{X^{(i)}} \int_{X_i} \int_{X_i}$$
$$F_0 \left(\left((x_{\omega'_j, j})_{j \in A \setminus \{i\}}, x_{0,i} \right) \right)_{\omega' \in A \setminus \{i\}}$$
$$\overline{F_1} \left(\left((x_{\omega'_j, j})_{j \in A \setminus \{i\}}, x_{1,i} \right) \right)_{\omega' \in A \setminus \{i\}}$$
$$d\mu_i(x_{0,i}) d\mu_i(x_{1,i}) d\mu^{(i)}(x_0^{(i)}) d\mu^{(i)}(x_1^{(i)})$$

for all $i \in A$ and $F_0, F_1 \in L^\infty(X)^{[A \setminus \{i\}]} \subset L^\infty(X^{\{0,1\}^{A \setminus \{i\}}})$, where $X^{(i)} := \prod_{j \in A \setminus \{i\}} X_j$, $\mu^{(i)} := \prod_{j \in A \setminus \{i\}} \mu_j$, and $x_a^{(i)} = (x_{a,j})_{j \in A \setminus \{i\}}$ for $a = 0, 1$. From this formula one can verify the inner product property without much trouble (the main difficulty here is simply in unpacking all the notation).

The third example is that of the (incomplete) *Gowers-Host-Kra spaces* $U^A \cap L^\infty(X)$. Here, $X = (X, \mathcal{B}, \mu)$ is a probability space with an invertible measure-preserving shift T, which of course induces a measure-preserving action $n \mapsto T^n$ of the integers $\mathbf{Z}$ on X. (One can replace the integers in the discussion that follows by more general nilpotent amenable groups, but we will stick to integer actions for simplicity.) It is often convenient to also assume that the measure μ is ergodic, though this is not strictly required to define the semi-norms. The space here is $L^\infty(X)$; the power $L^\infty(X)^{[A]}$ is then a subspace of $L^\infty(X^{\{0,1\}^A})$. One can define the *Host-Kra measure* $\mu^{[A]}$ on $X^{[A]}$ for any finite A by the following recursive procedure. First, when A is empty, then $\mu^{[A]}$ is just μ. If instead A is non-empty, then pick an element $i \in A$ and view $X^{[A]}$ as the Cartesian product of $X^{[A \setminus \{i\}]}$ with itself. The shift T acts on X, and thus acts diagonally on $X^{[A \setminus \{i\}]}$ by acting on each component separately. It is not hard to show inductively from the construction that we are about to give that $\mu^{[A \setminus \{i\}]}$ is invariant with respect to this diagonal shift, which we will call $T^{[A \setminus \{i\}]}$. The product σ-algebra $\mathcal{B}^{[A \setminus \{i\}]}$ has an invariant factor $(\mathcal{B}^{[A \setminus \{i\}]})^{T^{[A \setminus \{i\}]}}$ with respect to this shift. We then define $\mu^{[A]}$ to be the *relative* product of $\mu^{[A \setminus \{i\}]}$ with itself relative to this invariant factor. One can show that this definition is independent of the choice of i, and that the form

$$\langle F \rangle_A := \int_{X^{[A]}} F \, d\mu^{[A]}$$

is an inner product of order A; see [**HoKr2005**] for details.

A final (and significantly simpler) example of a inner product space of order A is the Lebesgue space $L^{2^{|A|}}(X)$ on some measure space $X = (X, \mathcal{B}, \mu)$, with inner product

$$\langle F \rangle_A := \int_X F((x, \ldots, x)) \, d\mu(x)$$

where $x \mapsto (x, \ldots, x)$ is the diagonal embedding from X to $X^{[A]} \equiv X^{2^{|A|}}$. For tensor products, this inner product takes the form

$$\langle \bigotimes_{\omega \in \{0,1\}^A} \mathcal{C}^{|\omega|} f_\omega \rangle_A = \int_X \prod_{\omega \in \{0,1\}^A} \mathcal{C}^{|\omega|} f_\omega \, d\mu;$$

thus, for instance, when $A = \{1, 2\}$,

$$\langle f_{00} \otimes \overline{f_{01}} \otimes \overline{f_{10}} \otimes f_{11} \rangle_A = \int_X f_{00} \overline{f_{01}} \overline{f_{10}} f_{11} \, d\mu.$$

We leave it as an exercise to the reader to show $L^{2^{|A|}}(X)$ is indeed an inner product space of order A. This example is (the completion of) the Gowers-Host-Kra space in the case when the shift T is trivial.

We also remark that given an inner product space $(V, \langle \rangle_A)$ of some order A, given some subset B of A, and given a fixed vector v_* in V, one can define a weighted inner product space $(V, \langle \rangle_{B,v_*})$ of order B by the formula

$$\langle F \rangle_{B, v_*} := \langle F \otimes \bigotimes_{\omega \in \{0,1\}^A \setminus \{0,1\}^B} \mathcal{C}^{|\omega|} v_* \rangle_A$$

for all $F \in V^{[B]}$, where $\{0,1\}^B$ is embedded in $\{0,1\}^A$ by extension by zero and the tensor product on the right-hand side is defined in the obvious manner. One can check that this is indeed a weighted inner product space. This is a generalisation of the classical fact that every vector v_* in an inner product space V naturally defines a linear functional $w \mapsto \langle w, v_* \rangle$ on V. In the case of the Gowers uniformity spaces with $v_* := 1$, this construction takes $U^A(G)$ to $U^B(G)$; similarly for the Gowers box spaces.

2.2.3. Basic theory. Let V be an inner product space of order A for some finite non-empty A. The splitting axiom tells us that

$$\langle F_0 \otimes_i \overline{F_1} \rangle_A = \langle F_0, F_1 \rangle_{A \setminus \{i\}}$$

for all $i \in A$, $F_0, F_1 \in V^{[A \setminus \{i\}]}$, and some inner product $\langle , \rangle$ on $X^{[A \setminus \{i\}]}$. In particular, one has

$$\langle F \otimes_i \overline{F} \rangle_A \geq 0$$

for all $F \in V^{[A \setminus \{i\}]}$, as well as the classical Cauchy-Schwarz inequality

$$|\langle F_0 \otimes_i \overline{F_1} \rangle_A| \leq |\langle F_0 \otimes_i \overline{F_0} \rangle_A|^{1/2} |\langle F_1 \otimes_i \overline{F_1} \rangle_A|^{1/2}.$$

2.2. Higher order Hilbert spaces

If we specialise this inequality to the tensor products

$$F_a := \bigotimes_{\omega' \in \{0,1\}^{A\setminus\{i\}}} \mathcal{C}^{|\omega'|} v_{a,\omega'}$$

for various $v_{a,\omega'} \in V$, one concludes that

$$|\langle \bigotimes_{\omega \in \{0,1\}^A} \mathcal{C}^{|\omega|} v_\omega \rangle_A| \leq \prod_{a \in \{0,1\}} |\langle \bigotimes_{\omega \in \{0,1\}^A} \mathcal{C}^{|\omega|} v_{a,\omega'} \rangle_A|^{1/2}$$

where we write $\omega = (\omega_i, \omega')$ for some $\omega_i \in \{0,1\}$ and $\omega' \in \{0,1\}^{A\setminus\{i\}}$. If we iterate this inequality once for each $i \in A$, we obtain the *Cauchy-Schwarz-Gowers inequality*

$$|\langle \bigotimes_{\omega \in \{0,1\}^A} \mathcal{C}^{|\omega|} v_\omega \rangle_A| \leq \prod_{\omega \in \{0,1\}^A} \|v_\omega\|_A$$

where

$$\|v\|_A := |\langle \bigotimes_{\omega \in \{0,1\}^A} \mathcal{C}^{|\omega|} v \rangle_A|^{1/2^{|A|}}.$$

The quantity $\|v\|_A$ is clearly non-negative and homogeneous. We also have the *Gowers triangle inequality*

$$\|v_0 + v_1\|_A \leq \|v_0\|_A + \|v_1\|_A,$$

which makes $\|\cdot\|_A$ a semi-norm (and in fact a norm, if the inner product space was positive definite). To see this inequality, we first raise both sides to the power $2^{|A|}$:

$$\|v_0 + v_1\|_A^{2^{|A|}} \leq (\|v_0\|_A + \|v_1\|_A)^{2^{|A|}}.$$

The left-hand side can be expanded as

$$|\langle \bigotimes_{\omega \in \{0,1\}^A} \mathcal{C}^{|\omega|}(v_0 + v_1) \rangle_A|$$

which after expanding out using linearity and the triangle inequality, can be bounded by

$$\sum_{\alpha \in \{0,1\}^{\{0,1\}^A}} |\langle \bigotimes_{\omega \in \{0,1\}^A} \mathcal{C}^{|\omega|} v_{\alpha_\omega} \rangle_A|$$

which by the Cauchy-Schwarz-Gowers inequality can be bounded in turn by

$$\sum_{\alpha \in \{0,1\}^{\{0,1\}^A}} \prod_{\omega \in \{0,1\}^A} \|v_{\alpha_\omega}\|_A$$

which can then be factored into $(\|v_0\|_A + \|v_1\|_A)^{2^{|A|}}$ as required.

Note that when A is a singleton set, the above argument collapses to the usual derivation of the triangle inequality from the classical Cauchy-Schwarz inequality. It is also instructive to see how this collapses to one of the standard proofs of the triangle inequality for $L^{2^k}(X)$ using a large number of applications of the Cauchy-Schwarz inequality.

In analogy with classical Hilbert spaces, one can define a *Hilbert space of order A* to be an inner product space V of order A which is both positive definite and complete, so that the norm $\|\|_A$ gives V the structure of a Banach space. A typical example is $U^2(G)$ for a finite abelian G, which is the space of all functions $f\colon G \to \mathbf{C}$ with the norm

$$\|f\|_{U^2(G)} = \|\hat{f}\|_{\ell^4(\hat{G})}$$

where $\hat{G}$ is the Pontraygin dual of G (i.e., the space of homomorphisms $\xi\colon x \mapsto \xi \cdot x$ from G to $\mathbf{R}/\mathbf{Z}$) and $\hat{f}(\xi) := \mathbf{E}_{x \in G} f(x) e(-\xi \cdot x)$ is the Fourier transform. Thus we see that $\ell^4(\hat{G})$ is a Hilbert space of order 2. More generally, $L^{2^k}(X)$ for any measure space X and any $k \geq 0$ can be viewed as a Hilbert space of order k.

The Gowers norms $U^d(G)$ and Gowers-Host-Kra norms $U^d(X)$ coincide in the model case when $X = G = \mathbf{Z}/N\mathbf{Z}$ is a cyclic group with uniform measure and the standard shift $T\colon x \mapsto x + 1$. Also, the Gowers norms $U^d(G)$ can be viewed as a special case of the box norms via the identity

$$\|f\|_{U^d(G)} := \|f \circ s\|_{\Box^d(G^d)}$$

where $s\colon G^d \to G$ is the summation operation $s(x_1, \ldots, x_d) := x_1 + \cdots + x_d$.

Just as classical inner product spaces can be made positive definite by quotienting out the norm zero elements, and then made into a classical Hilbert space by metric completion, inner product spaces of any order can also be made positive definite and completed. One can apply this procedure for instance to obtain the completed Gowers box spaces $\Box^A(X)$ and the completed Gowers-Host-Kra spaces $U^A(X)$ (which become $L^{2^{|A|}}(X)$ when the shift T is trivial). These spaces are related, but not equal, to their Lebesgue counterparts $L^p(X)$; for instance, for the Gowers-Host-Kra spaces in the ergodic setting, a repeated application of Young's inequality reveals the inequalities

$$\|f\|_{U^A(X)} \leq \|f\|_{L^{2^{|A|}/(|A|+1)}(X)} \leq \|f\|_{L^\infty(X)},$$

and so $U^A(X)$ contains a (quotient) of $L^{2^{|A|}/(|A|+1)}(X)$.

The null space of the Gowers-Host-Kra norm $U^A(X)$ in $L^\infty(X)$ in the ergodic case is quite interesting; it turns out to be the space $L^\infty(\mathcal{Z}_{<|A|})^\perp$ of bounded measurable functions f whose conditional expectation $\mathbf{E}(f|\mathcal{Z}_{<|A|})$ on the *characteristic factor* $\mathcal{Z}_{<|A|}$ of order $|A|-1$ of X vanishes; in particular, $L^\infty(\mathcal{Z}_{<|A|})$ becomes a dense subspace of $U^A(X)$, embedded injectively. It is a highly non-trivial and useful result, first obtained in [**HoKr2005**], that

2.2. Higher order Hilbert spaces

$\mathcal{Z}_{<|A|}$ is the inverse limit of all nilsystem factors of step at most $|A|-1$; this is the ergodic counterpart of the *inverse conjecture for the Gowers norms*.

2.2.4. The category of higher order inner product spaces. The higher order Hilbert spaces $L^1(X), L^2(X), L^4(X), L^8(X), \ldots$ are related to each other via Hölder's inequality; the pointwise product of two L^4 functions is in L^2, the product of two L^8 functions is in L^4, and so forth. Furthermore, the inner products on all of these spaces are can be connected to each other via the pointwise product.

We can generalise this concept, giving the class of inner product spaces (of arbitrary orders) the structure of a category.

Definition 2.2.2. Let $B \subseteq A$ be finite sets, and let $V_B = (V_B, \langle \rangle_B)$, $V_A = (V_A, \langle \rangle_A)$ be inner product spaces of order B, A, respectively. An *isometry* ϕ from V_A to V_B is a linear map

$$\phi: \bigotimes_{\omega \in \{0,1\}^{A \setminus B}} \mathcal{C}^{|\omega|} V_A \to V_B$$

which preserves the inner product in the sense that

$$\left\langle \bigotimes_{\omega \in \{0,1\}^A} \mathcal{C}^{|\omega|} v_\omega \right\rangle_A = \left\langle \bigotimes_{\omega' \in \{0,1\}^B} \mathcal{C}^{|\omega'|} \phi\left(\bigotimes_{\omega'' \in \{0,1\}^{A \setminus B}} \mathcal{C}^{|\omega''|} v_{(\omega',\omega'')} \right) \right\rangle_B,$$

where $\omega', \omega'' \to (\omega', \omega'')$ is the obvious concatenation map from $\{0,1\}^B \times \{0,1\}^{A \setminus B}$ to $\{0,1\}^A$.

Given an isometry ϕ from V_A to V_B, and an isometry ψ from V_B to V_C for some $C \subset B \subset A$, one can form the composition

$$\psi \circ \phi: \bigotimes_{\omega \in \{0,1\}^{A \setminus C}} \mathcal{C}^{|\omega|} V_A \to V_C$$

by the formula

$$\psi \circ \phi \left(\bigotimes_{\omega \in \{0,1\}^{A \setminus C}} \mathcal{C}^{|\omega|} v_\omega \right)$$

$$:= \psi \left(\bigotimes_{\omega' \in \{0,1\}^{B \setminus C}} \mathcal{C}^{|\omega'|} \phi \left(\bigotimes_{\omega'' \in \{0,1\}^{A \setminus B}} \mathcal{C}^{|\omega''|} v_{(\omega',\omega'')} \right) \right)$$

and extending by linearity; one can verify that this continues to be an isometry, and that the class of inner product spaces of arbitrary order together with isomorphisms form a category.

When $A = B$ is a singleton set, the above concept collapses to the classical notion of an isometry for inner product spaces. Of course, one could specialise to the subcategory of higher order Hilbert spaces if desired. The inner product on a higher order inner product space can now be interpreted as an isometry from that space to the space $\mathbf{C}$ (viewed as an inner product space of order $\emptyset$), and is the unique such isometry; in the language of category theory, this space $\mathbf{C}$ becomes the *terminal object* of the category.

A model example of an isometry is the sesquilinear product map $f, g \mapsto f\bar{g}$, which is an isometry from $L^{2^d}(X)$ to $L^{2^{d-1}}(X)$ for any $d \geq 1$. For the Gowers-Host-Kra norms, the map $f, g \mapsto f \otimes \bar{g}$ is an isometry from $U^d(X^{[k]})$ to $U^{d-1}(X^{[k+1]})$ for any $d \geq 2$ and $k \geq 0$.

To see analogous isometries for the Gowers uniformity norms, one has to generalise these norms to the "non-ergodic" setting when one does not average the shift parameter h over the entire group G, but on a subgroup H. Specifically, for finite additive groups $H \leq G$ and functions $f_\omega : G \to \mathbf{C}$ with $\omega \in \{0,1\}^A$, define the *local Gowers inner product*

$$\langle \bigotimes_{\omega \in \{0,1\}^A} \mathcal{C}^{|\omega|} f_\omega \rangle_{U^A(G,H)} = \mathbf{E}_{x \in G, h_i \in H \forall i \in A} \prod_{\omega \in \{0,1\}^A} \mathcal{C}^{|\omega|} f_\omega(x + \sum_{i=1}^{A} \omega_i h_i).$$

By foliating G into cosets of H, one can express this local Gowers inner product as an amalgam of the ordinary Gowers inner product and a Lebesgue inner product. For instance, one has the identity

$$\|f\|_{U^A(G,H)} = \left(\sum_{y \in G/H} \|f(\cdot + y)\|_{U^A(H)}^{2^{|A|}} \right)^{1/2^{|A|}}.$$

We define the inner product space $U^A(G, H)$ to be the space of functions from G to $\mathbf{C}$ with the above inner product. Given any $j \in A$, we can then create an isometry $\Delta = \Delta_j$ from $U^A(G, H)$ to $U^{A\setminus\{j\}}(G \times H, H)$ by defining[14]

$$\Delta(f, f')(x, h) := f(x+h)\overline{f'(x)}.$$

One can obtain analogous isometries for the Gowers box norms after similarly generalising to "non-ergodic" settings; we leave this as an exercise to the interested reader.

Actually, the "derivative maps" from inner product spaces V_A of order A to those of order $A \setminus \{j\}$ can be constructed abstractly. Indeed, one can view $V_A \otimes \overline{V_A}$ as an inner product space of order $A \setminus \{j\}$ with the inner product

---

[14]This isometry does not ostensibly depend on j, except through the labels of the inner product of the target space $U^{A\setminus\{j\}}(G \times H, H)$ of the isometry.

2.2. Higher order Hilbert spaces

defined on tensor products by

$$\left\langle \bigotimes_{\omega' \in \{0,1\}^{A \setminus \{j\}}} \mathcal{C}^{|\omega'|}(v_{\omega',0} \otimes \overline{v_{\omega',1}}) \right\rangle_{A \setminus \{j\}}$$
$$:= \left\langle \bigotimes_{(\omega',\omega_j) \in \{0,1\}^A} \mathcal{C}^{|(\omega',\omega_j)|} v_{\omega',\omega_j} \right\rangle_A$$

and then the map $v, w \mapsto v \otimes \overline{w}$ is an isometry. One can iterate this construction and obtain a *cubic complex* of inner product spaces

$$V_B := \bigotimes_{\omega \in \{0,1\}^{A \setminus B}} \mathcal{C}^{|\omega|} V_A$$

of order B for each $B \subset A$, together with a commuting system of derivative isometries Δ from V_B to $V_{B \setminus \{j\}}$ for each $j \in B \subset A$.

Conversely, one can use cubic complexes to build higher order inner product spaces:

Proposition 2.2.3. *Let A be a finite set. For each $B \subset A$, suppose that we have a vector space V_B equipped with a $\{0,1\}^B$-sesquilinear form*

$$\langle \rangle_B : \bigotimes_{\omega \in \{0,1\}^B} \mathcal{C}^{|\omega|} V_B \to \mathbf{C}$$

and suppose that for each $j \in B$ one has a sesquilinear product

$$\Delta_{B \to B \setminus \{j\}} : V_B \otimes \overline{V_B} \to V_{B \setminus \{j\}}$$

obeying the compatibility conditions

$$\left\langle \bigotimes_{\omega \in \{0,1\}^B} \mathcal{C}^{|\omega|} v_\omega \right\rangle_B$$
$$= \left\langle \bigotimes_{\omega' \in \{0,1\}^{B \setminus \{j\}}} \Delta_{B \to B \setminus \{j\}}(v_{(\omega',0)}, v_{(\omega',1)}) \right\rangle_{B \setminus \{j\}}$$

whenever $v_\omega \in V_B$ for all $\omega \in \{0,1\}^B$. Suppose also that the form $\langle \rangle_{\{j\}}$ is a classical inner product on $V_{\{j\}}$ for every $j \in A$. Then for each $B \subset A$, V_B is an inner product space of order j, and the maps $\Delta_{B \to B \setminus \{j\}}$ become isometries.

This proposition is established by an easy induction on the cardinality of B. Note that we do not require the derivative maps $\Delta_{B \to B \setminus \{j\}}$ to commute with each other, although this is almost always the case in applications.

2.3. The uncertainty principle

A recurring theme in mathematics is that of *duality*: a mathematical object X can either be described *internally* (or in *physical space*, or *locally*), by describing what X physically consists of (or what kind of maps exist *into* X), or *externally* (or in *frequency space*, or *globally*), by describing what X globally interacts or resonates with (or what kind of maps exist *out of* X). These two fundamentally opposed perspectives on the object X are often *dual* to each other in various ways: performing an operation on X may transform it one way in physical space, but in a dual way in frequency space, with the frequency space description often being an "inversion" of the physical space description. In several important cases, one is fortunate enough to have some sort of *fundamental theorem* connecting the internal and external perspectives. Here are some (closely inter-related) examples of this perspective:

(i) **Vector space duality.** A vector space V over a field F can be described either by the set of vectors inside V, or dually by the set of linear functionals $\lambda\colon V \to F$ from V to the field F (or equivalently, the set of vectors inside the dual space V^*). (If one is working in the category of topological vector spaces, one would work instead with continuous linear functionals; and so forth.) A fundamental connection between the two is given by the *Hahn-Banach theorem* (and its relatives); see e.g. [**Ta2010**, §1.5].

(ii) **Vector subspace duality.** In a similar spirit, a subspace W of V can be described either by listing a basis or a spanning set, or dually by a list of linear functionals that cut out that subspace (i.e. a spanning set for the orthogonal complement $W^\perp := \{\lambda \in V^* : \lambda(w) = 0 \text{ for all } w \in W\}$). Again, the Hahn-Banach theorem provides a fundamental connection between the two perspectives.

(iii) **Convex duality.** More generally, a (closed, bounded) convex body K in a vector space V can be described either by listing a set of (extreme) points whose convex hull is K, or else by listing a set of (irreducible) linear inequalities that cut out K. The fundamental connection between the two is given by the *Farkas lemma*; see [**Ta2008**, §1.16] for further discussion.

(iv) **Ideal-variety duality.** In a slightly different direction, an algebraic variety V in an affine space A^n can be viewed either "in physical space" or "internally" as a collection of points in V, or else "in frequency space" or "externally" as a collection of polynomials on A^n whose simultaneous zero locus cuts out V. The fundamental

2.3. The uncertainty principle

connection between the two perspectives is given by the *nullstellensatz*, which then leads to many of the basic fundamental theorems in classical algebraic geometry; see [**Ta2008**, §1.15] for further discussion.

(v) **Hilbert space duality.** An element v in a Hilbert space H can either be thought of in physical space as a *vector* in that space, or in momentum space as a *covector* $w \mapsto \langle v, w \rangle$ on that space. The fundamental connection between the two is given by the *Riesz representation theorem for Hilbert spaces*; see [**Ta2010**, §1.15] for further discussion.

(vi) **Semantic-syntactic duality** Much more generally still, a mathematical theory can either be described *internally* or *syntactically* via its axioms and theorems, or *externally* or *semantically* via its models. The fundamental connection between the two perspectives is given by the *Gödel completeness theorem*; see [**Ta2010b**, §1.4] for further discussion.

(vii) **Intrinsic-extrinsic duality.** A (Riemannian) manifold M can either be viewed intrinsically (using only concepts that do not require an ambient space, such as the *Levi-Civita connection*), or extrinsically, for instance, as the level set of some defining function in an ambient space. Some important connections between the two perspectives includes the *Nash embedding theorem* and the *theorema egregium*.

(viii) **Group duality.** A group G can be described either via *presentations* (lists of generators, together with relations between them) or *representations* (realisations of that group in some more concrete group of transformations). A fundamental connection between the two is *Cayley's theorem*. Unfortunately, in general it is difficult to build upon this connection (except in special cases, such as the abelian case), and one cannot always pass effortlessly from one perspective to the other.

(ix) **Pontryagin group duality.** A (locally compact Hausdorff) *abelian* group G can be described either by listing its elements $g \in G$, or by listing the *characters* $\chi \colon G \to \mathbf{R}/\mathbf{Z}$ (i.e., continuous homomorphisms from G to the unit circle, or equivalently elements of $\hat{G}$). The connection between the two is the focus of *abstract harmonic analysis*; see [**Ta2010**, §1.12] for further discussion.

(x) **Pontryagin subgroup duality.** A subgroup H of a locally compact abelian group G can be described either by generators in H, or generators in the orthogonal complement $H^\perp := \{\xi \in \hat{G} : \xi \cdot h =$

0 for all $h \in H$}. One of the fundamental connections between the two is the *Poisson summation formula*.

(xi) **Fourier duality.** A (sufficiently nice) function $f \colon G \to \mathbf{C}$ on a locally compact abelian group G (equipped with a Haar measure μ) can either be described in physical space (by its values $f(x)$ at each element x of G) or in frequency space (by the values $\hat{f}(\xi) = \int_G f(x)e(-\xi \cdot x)\, d\mu(x)$ at elements ξ of the Pontryagin dual $\hat{G}$). The fundamental connection between the two is the *Fourier inversion formula*.

(xii) **The uncertainty principle.** The behaviour of a function f at physical scales above (resp. below) a certain scale R is almost completely controlled by the behaviour of its Fourier transform $\hat{f}$ at frequency scales below (resp. above) the dual scale $1/R$ and vice versa, thanks to various mathematical manifestations[15] of the *uncertainty principle*.

(xiii) **Stone/Gelfand duality.** A (locally compact Hausdorff) topological space X can be viewed in physical space (as a collection of points), or dually, via the C^* algebra $C(X)$ of continuous complex-valued functions on that space, or (in the case when X is compact and totally disconnected) via the Boolean algebra of clopen sets (or equivalently, the idempotents of $C(X)$). The fundamental connection between the two is given by the *Stone representation theorem* (see [**Ta2010**, §2.3]) or the (commutative) *Gelfand-Naimark theorem* (see [**Ta2010**, §1.10]).

In this section we will discuss one particular manifestation of duality, namely the *uncertainty principle* that describes the dual relationship between physical space and frequency space. There are various concrete formalisations of this principle, most famously the *Heisenberg uncertainty principle* and the *Hardy uncertainty principle* (see [**Ta2010**, §2.6]); but in many situations, it is the *heuristic* formulation of the principle that is more useful and insightful than any particular rigorous theorem that attempts to capture that principle. Unfortunately, it is a bit tricky to formulate this heuristic in a succinct way that covers all the various applications of that principle; the Heisenberg inequality $\Delta x \cdot \Delta \xi \gtrsim 1$ is a good start, but it only captures a portion of what the principle tells us. Consider, for instance, the following (deliberately vague) statements, each of which can be viewed (heuristically, at least) as a manifestation of the uncertainty principle:

---

[15] The Poisson summation formula can also be viewed as a variant of this principle, using subgroups instead of scales.

2.3. The uncertainty principle

(i) A function which is band-limited (restricted to low frequencies) is featureless and smooth at fine scales, but can be oscillatory (i.e. containing plenty of cancellation) at coarse scales. Conversely, a function which is smooth at fine scales will be almost entirely restricted to low frequencies.

(ii) A function which is restricted to high frequencies is oscillatory at fine scales, but is negligible at coarse scales. Conversely, a function which is oscillatory at fine scales will be almost entirely restricted to high frequencies.

(iii) Projecting a function to low frequencies corresponds to averaging out (or spreading out) that function at fine scales, leaving only the coarse scale behaviour.

(iv) Projecting a frequency to high frequencies corresponds to removing the averaged coarse scale behaviour, leaving only the fine scale oscillation.

(v) The number of degrees of freedom of a function is bounded by the product of its spatial uncertainty and its frequency uncertainty (or more generally, by the volume of the phase space uncertainty). In particular, there are not enough degrees of freedom for a non-trivial function to be simulatenously localised to both very fine scales and very low frequencies.

(vi) To control the coarse scale (or global) averaged behaviour of a function, one essentially only needs to know the low frequency components of the function (and vice versa).

(vii) To control the fine scale (or local) oscillation of a function, one only needs to know the high frequency components of the function (and vice versa).

(viii) Localising a function to a region of physical space will cause its Fourier transform (or inverse Fourier transform) to resemble a plane wave on every dual region of frequency space.

(ix) Averaging a function along certain spatial directions or at certain scales will cause the Fourier transform to become localised to the dual directions and scales. The smoother the averaging, the sharper the localisation.

(x) The smoother a function is, the more rapidly decreasing its Fourier transform (or inverse Fourier transform) is (and vice versa).

(xi) If a function is smooth or almost constant in certain directions or at certain scales, then its Fourier transform (or inverse Fourier transform) will decay away from the dual directions or beyond the dual scales.

(xii) If a function has a singularity spanning certain directions or certain scales, then its Fourier transform (or inverse Fourier transform) will decay slowly along the dual directions or within the dual scales.

(xiii) Localisation operations in position approximately commute with localisation operations in frequency so long as the product of the spatial uncertainty and the frequency uncertainty is significantly larger than one.

(xiv) In the high frequency (or large scale) limit, position and frequency asymptotically behave like a pair of classical observables, and partial differential equations asymptotically behave like classical ordinary differential equations. At lower frequencies (or finer scales), the former becomes a "quantum mechanical perturbation" of the latter, with the strength of the quantum effects increasing as one moves to increasingly lower frequencies and finer spatial scales.

(xv) Etc., etc.

(xvi) Almost all of the above statements generalise to other locally compact abelian groups than $\mathbf{R}$ or $\mathbf{R}^n$, in which the concept of a direction or scale is replaced by that of a subgroup or an approximate subgroup[16].

All of the above (closely related) assertions can be viewed as being instances of "the uncertainty principle", but it seems difficult to combine them all into a single unified assertion, even at the heuristic level; they seem to be better arranged as a *cloud* of tightly interconnected assertions, each of which is reinforced by several of the others. The famous inequality $\Delta x \cdot \Delta \xi \gtrsim 1$ is at the centre of this cloud, but is by no means the only aspect of it.

The uncertainty principle (as interpreted in the above broad sense) is one of the most fundamental principles in harmonic analysis (and more specifically, to the subfield of *time-frequency analysis*), second only to the Fourier inversion formula (and more generally, *Plancherel's theorem*) in importance; understanding this principle is a key piece of intuition in the subject that one has to internalise before one can really get to grips with this subject (and also with closely related subjects, such as semi-classical analysis and microlocal analysis). Like many fundamental results in mathematics, the principle is not actually that difficult to understand, once one sees how it works; and when one needs to use it rigorously, it is usually not too difficult to improvise a suitable formalisation of the principle for the occasion. But, given how vague this principle is, it is difficult to present this principle in a traditional "theorem-proof-remark" manner. Here, we will give a set of interrelated discussions about this principle rather than a linear development

---

[16] In particular, as we will see below, the Poisson summation formula can be viewed as another manifestation of the uncertainty principle.

2.3. The uncertainty principle

of the theory, as this seemed to more closely align with the nature of this principle.

The uncertainty principle given here is associated only to classical (or *linear*) Fourier analysis. In principle, there should be uncertainty principles for quadratic or higher order Fourier analysis, but we will not pursue such questions here.

2.3.1. An informal foundation for the uncertainty principle. Many of the manifestations of the uncertainty principle can be heuristically derived from the following informal heuristic:

Heuristic 2.3.1 (Phase heuristic)**.** *If the phase $\phi(x)$ of a complex exponential $e^{2\pi i \phi(x)}$ fluctuates by less than 1 for x in some nice domain Ω (e.g. a convex set, or more generally an approximate subgroup), then the phase $e^{2\pi i \phi(x)}$ behaves as if it were constant on Ω. If instead the phase fluctuates by much more than 1, then $e^{2\pi i \phi(x)}$ should oscillate and exhibit significant cancellation. The more the phase fluctuates, the more oscillation and cancellation becomes present.*

For instance, according to this heuristic, on an interval $[-R, R]$ in the real line, the linear phase $x \mapsto e^{2\pi i \xi x}$ at a given frequency $\xi \in \mathbf{R}$ behaves like a constant when $|\xi| \ll 1/R$, but oscillates significantly when $|\xi| \gg 1/R$. This is visually plausible if one graphs the real and imaginary parts $\cos(2\pi i \xi x)$, $\sin(2\pi i \xi x)$. For now, we will take this principle as axiomatic, without further justification, and without further elaboration as to what vague terms such as "behaves as if" or $\ll$ mean.

Remark 2.3.2. The above heuristic can also be viewed as the informal foundation for the *principle of stationary phase*. This is not coincidental, but will not be the focus of the discussion here.

Let's give a few examples to illustrate how this heuristic informally implies some versions of the uncertainty principle. Suppose, for instance, that a function $f \colon \mathbf{R} \to \mathbf{C}$ is supported in an interval $[-R, R]$. Now consider the Fourier transform[17]

$$\hat{f}(\xi) := \int_{\mathbf{R}} e^{-2\pi i x \xi} f(x) \, dx = \int_{-R}^{R} e^{-2\pi i x \xi} f(x) \, dx.$$

We assume that the function is nice enough (e.g. absolutely integrable will certainly suffice) that one can define the Fourier transform without difficulty.

If $|\xi| \ll 1/R$, then the phase $x\xi$ fluctuates by less than 1 on the domain $x \in [-R, R]$, and so the phase here is essentially constant by the above

---

[17] Other normalisations of the Fourier transform are also used in the literature, but the precise choice of normalisation does not significantly affect the discussion here.

heuristic; in particular, we expect the Fourier transform $\hat{f}(\xi)$ to not vary much in this interval. More generally, if we consider frequencies ξ in an interval $|\xi - \xi_0| \ll 1/R$ for a fixed ξ_0, then on separating $e^{-2\pi i x \xi}$ as $e^{-2\pi i x \xi_0} \times e^{-2\pi i x (\xi - \xi_0)}$, the latter phase $x(\xi - \xi_0)$ is essentially constant by the above heuristic, and so we expect $\hat{f}(\xi)$ to not vary much in this interval either. Thus $\hat{f}(\xi)$ is close to constant at scales much finer than $1/R$, just as the uncertainty principle predicts.

A similar heuristic calculation using the Fourier inversion formula

$$f(x) = \int_{\mathbf{R}} e^{2\pi i x \xi} \hat{f}(\xi) \, d\xi$$

shows that if the Fourier transform $\hat{f}(\xi)$ is restricted to an interval $[-N, N]$, then the function f should behave roughly like a constant at scales $\ll 1/N$. A bit more generally, if the Fourier transform is restricted to an interval $[\xi_0 - N, \xi_0 + N]$, then by separating $e^{2\pi i x \xi}$ as $e^{2\pi i x_0 \xi_0} e^{2\pi i (x - x_0) \xi} e^{2\pi i x_0 (\xi - \xi_0)} e^{2\pi i (x - x_0)(\xi - \xi_0)}$ and discarding the last phase when $|x - x_0| \ll 1/N$, we see that the function f behaves like a constant multiple of the plane wave $x \mapsto e^{2\pi i x \xi_0}$ on each interval $\{x : |x - x_0| \ll 1/N\}$ (but it could be a different constant multiple on each such interval).

The same type of heuristic computation can be carried through in higher dimensions. For instance, if a function $f \colon \mathbf{R}^n \to \mathbf{C}$ has Fourier transform supported in some symmetric convex body Ω, then one expects f itself to behave like a constant on any translate $x_0 + c\Omega^*$ of a small multiple $0 < c \ll 1$ of the *polar body*

$$\Omega^* := \{x \in \mathbf{R}^n : |x \cdot \xi| \leq 1 \text{ for all } \xi \in \Omega\}$$

of Ω.

An important special case where the above heuristics are in fact exactly rigorous is when one does not work with *approximate subgroups* such as intervals $[-R, R]$ or convex bodies Ω, but rather with *subgroups* H of the ambient (locally compact abelian) group G that is serving as physical space. Here, of course, we need the general Fourier transform

$$\hat{f}(\xi) := \int_G e^{-2\pi i \xi \cdot x} f(x) \, d\mu_G(x),$$

where μ_G is a Haar measure on the locally compact abelian group G, where $\xi \colon x \mapsto \xi \cdot x$ is a continuous homomorphism from G to $\mathbf{R}/\mathbf{Z}$ (and is thus an element of the Pontryagin dual group $\hat{G}$), with Fourier transform given by the inversion formula

$$f(x) = \int_{\hat{G}} e^{2\pi i \xi \cdot x} \hat{f}(\xi) d\mu_{\hat{G}}(\xi)$$

2.3. The uncertainty principle

where $\mu_{\hat{G}}$ is the dual Haar measure on $\hat{G}$ (see, e.g., my lecture notes for further discussion of this general theory). If f is supported on a subgroup H of G (this may require f to be a measure rather than a function, if H is a measure zero subgroup of G), we conclude[18] (rigorously!) that $\hat{f}$ is constant along cosets of the orthogonal complement

$$\hat{H} := \{\xi \in \hat{G} : \xi \cdot x = 0 \text{ for all } x \in H\}.$$

For instance, a measure f on $\mathbf{R}$ that is supported on $\mathbf{Z}$ will have a Fourier transform $\hat{f}$ that is constant along the $\mathbf{Z}$ direction, as $\mathbf{Z}$ is its own orthogonal complement. This is a basic component of the Poisson summation formula.

Remark 2.3.3. Of course, in Euclidean domains such as $\mathbf{R}$ or $\mathbf{R}^n$, basic sets such as the intervals $[-R, R]$ are not actual subgroups, but are only *approximate subgroups* (roughly speaking, this means that they are closed under addition a "reasonable fraction of the time"; for a precise definition, see [**TaVu2006**]. However, there are *dyadic models* of Euclidean domains (cf. [**Ta2008**, §1.6]), such as the field $F((\frac{1}{t}))$ of formal Laurent series in a variable $\frac{1}{t}$ over a finite field F, in which the analogues of such intervals *are* in fact actual subgroups, which allows for a very precise and rigorous formalisation of many of the heuristics given here in that setting.

One can view an interval such as $[-1/R, 1/R]$ as being an approximate orthogonal complement to the interval $[-R, R]$, and more generally the polar body Ω^* as an approximate orthogonal complement to Ω. Conversely, the uncertainty principle $\Delta x \cdot \Delta \xi \gg 1$ when specialised to subgroups H of a finite abelian group G becomes the equality

$$|H| \cdot |H^\perp| = |G|$$

and when specialised to subspaces V of a Euclidean space $\mathbf{R}^n$ becomes

$$\dim(V) + \dim(V^\perp) = \dim(\mathbf{R}^n).$$

We saw above that a function f that was restricted to a region Ω would necessarily have a Fourier transform $\hat{f}$ that was essentially constant on translates of (small multiples of) the dual region Ω^*. This implication can be partially reversed. For instance, suppose that $\hat{f}$ behaved like a constant at all scales $\ll N$. Then if one inspects the Fourier inversion formula

$$f(x) = \int_{\mathbf{R}} \hat{f}(\xi) e^{2\pi i x \xi} \, d\xi$$

we note that if $|x| \gg 1/N$, then $e^{2\pi i x \xi}$ oscillates at scales $\ll N$ by the above heuristic, and so $f(x)$ should be negligible when $|x| \gg 1/N$.

---

[18] This is assuming that f is a function or a measure. If f is merely a distribution, the situation is more complicated.

The above heuristic computations can be made rigorous in a number of ways. One basic method is to exploit the fundamental fact that the Fourier transform intertwines multiplication and convolution, thus[19]

$$\widehat{f * g} = \hat{f}\hat{g}$$

and

$$\widehat{fg} = \hat{f} * \hat{g}$$

and similarly for the inverse Fourier transform. For instance, if a function f has Fourier transform supported on $[-N, N]$, then we have

$$\hat{f} = \hat{f}\psi_N$$

where $\psi_N(x) := \psi(x/N)$ and ψ is a smooth and compactly supported (or rapidly decreasing) cutoff function[20] that equals 1 on the interval $[-1, 1]$.

Inverting the Fourier transform, we obtain the *reproducing formula*

$$f = f * \check{\psi}_N$$

where $\check{\psi}_N$ is the inverse Fourier transform of ψ_N. One can compute that

$$\check{\psi}_N(x) = N\check{\psi}(Nx)$$

and thus

(2.2) $$f(x) = \int_{\mathbf{R}} f(x + \frac{y}{N})\check{\psi}(y) \, dy.$$

If one chose ψ to be smooth and compactly supported (or at the very least, a Schwartz function), $\check{\psi}$ will be in the Schwartz class. As such, (2.2) can be viewed as an assertion that the value of the band-limited function f at any given point x is essentially an average of its values at nearby points $x + \frac{y}{N}$ for $y = O(1)$. This formula can already be used to give many rigorous instantiations of the uncertainty principle.

Remark 2.3.4. Another basic method to formalise the above heuristics, particularly with regard to "oscillation causes cancellation", is to use integration by parts.

---

[19]Here, the convolution $*$ is with respect to either the Haar measure μ_G on the physical space G, or the Haar measure $\mu_{\hat{G}}$ on the frequency space $\hat{G}$, as indicated by context.

[20]There is a lot of freedom here in what cutoff function to pick, but in practice, "all bump functions are usually equivalent"; unless one is optimising constants, needs a very specific and delicate cancellation, or if one really, really needs an explicit formula, one usually does not have to think too hard regarding what specific cutoff to use, though smooth and well localised cutoffs often tend to be superior to rough or slowly decaying cutoffs.

2.3.2. Projections. The restriction $1_{[-N,N]}(X)f := f1_{[-N,N]}$ of a function $f\colon \mathbf{R} \to \mathbf{C}$ to an interval $[-N,N]$ is just the orthogonal projection (in the Hilbert space $L^2(\mathbf{R})$) of f to the space of functions that are spatially supported in $[-N,N]$. Taking Fourier transforms (which, by Plancherel's theorem, preserves the Hilbert space $L^2(\mathbf{R})$), we see that the Fourier restriction $1_{[-N,N]}(D)f$ of f, defined as

$$\widehat{1_{[-N,N]}(D)f} := \hat{f} 1_{[-N,N]}$$

is the orthogonal projection of f to those functions with *Fourier* support in $[-N,N]$. As discussed above, such functions are (heuristically) those functions which are essentially constant at scales $\ll 1/N$. As such, these projection operators should behave like averaging operators at this scale. This turns out not to be that accurate of a heuristic if one uses the sharp cutoffs $1_{[-N,N]}$ (though this does work perfectly in the dyadic model setting), but if one replaces the sharp cutoffs by smoother ones, then this heuristic can be justified by using convolutions as in the previous section; this leads to *Littlewood-Paley theory*, a cornerstone of the harmonic analysis of function spaces such as Sobolev spaces, and which are particularly important in partial differential equations; see, for instance, [**Ta2006b**, Appendix A] for further discussion.

One can view the restriction operator $1_{[-N,N]}(X)$ as the spectral projection of the position operator $Xf(x) := xf(x)$ to the interval $[-N,N]$; in a similar vein, one can view $1_{[-N,N]}(D)$ as a spectral projection of the differentiation operator $Df(x) := \frac{1}{2\pi i}\frac{d}{dx}f(x)$.

As before, one can work with other sets than intervals here. For instance, restricting a function $f\colon G \to \mathbf{C}$ to a subgroup H causes the Fourier transform $\hat{f}$ to be averaged along the dual group $\hat{H}$. In particular, restricting a function $f\colon \mathbf{R} \to \mathbf{C}$ to the integers (and renormalising it to become the measure $\sum_{n\in\mathbf{Z}} f(n)\delta_n$) causes the Fourier transform $\hat{f}\colon \mathbf{R} \to \mathbf{C}$ to become summed over the dual group $\mathbf{Z}^\perp = \mathbf{Z}$ to become the function $\sum_{m\in\mathbf{Z}} \hat{f}(\cdot+m)$. In particular, the zero Fourier coefficient of $\sum_{n\in\mathbf{Z}} f(n)\delta_n$ is $\sum_{m\in\mathbf{Z}} \hat{f}(m)$, leading to the Poisson summation formula

$$\sum_{n\in\mathbf{Z}} f(n) = \sum_{m\in\mathbf{Z}} \hat{f}(m).$$

More generally, one has

$$\sum_{n\in R\mathbf{Z}} f(n) = \frac{1}{R} \sum_{m\in \frac{1}{R}\mathbf{Z}} \hat{f}(m)$$

for any $R > 0$, which can be viewed, on one hand, as a one-parameter family of identities interpolating between the inversion formula

$$f(0) = \int_{\mathbf{R}} \hat{f}(\xi) \, d\xi,$$

and, on the other hand, the forward Fourier transform formula

$$\int_{\mathbf{R}} f(x) \, dx = \hat{f}(0).$$

The duality $\Delta x \cdot \Delta \xi \gg 1$ between the position variable x and the frequency variable ξ (or equivalently, between the position operator X and the differentiation operator D) can be generalised to contexts in which the two dual variables have a different "physical" interpretation than position and frequency. One basic example of this is the duality $\Delta t \cdot \Delta E \gg 1$ between a time variable t and an energy variable E in quantum mechanics. Consider a time-dependent Schrödinger equation,

(2.3) $$i\partial_t \psi = H\psi; \quad \psi(0) = \psi_0,$$

for some Hermitian (and time-independent) spatial operator H on some arbitrary domain (which does not need to be a Euclidean space $\mathbf{R}^n$, or even a group), where we have normalised away for now the role of Planck's constant $\hbar$. If the underlying spatial space $L^2(\mathbf{R})$ has an orthonormal basis of eigenvector solutions to the time-independent Schrödinger equation

$$H u_k = E_k u_k,$$

then the solution to (2.3) is formally given by the formula

$$\psi = e^{-itH} \psi_0 = \sum_k e^{-iE_k t} \langle \psi_0, u_k \rangle u_k.$$

We thus see that the coefficients $\langle \psi_0, u_k \rangle$ (or more precisely, the eigenvectors $\langle \psi_0, u_k \rangle u_k$) can be viewed as the Fourier coefficients of ψ in time, with the energies E_k playing the role of the frequency vector. Taking traces, one (formally) sees a similar Fourier relationship between the trace function $\operatorname{tr}(e^{-itH})$ and the spectrum $E_1 < E_2 < E_3 < \ldots$:

(2.4) $$\operatorname{tr}(e^{-itH}) = \sum_k e^{-iE_k t}.$$

Here as a consequence, the heuristics of the uncertainty principle carry through. Just as the behaviour of a function f at scales $\ll T$ largely controls the spectral behaviour of $\hat{f}$ at scales $\gg 1/T$, one can use the evolution operator e^{-itH} of the Schrödinger equation up to times $|t| \leq T$ to understand

2.3. The uncertainty principle

the spectrum $E_1 < E_2 < E_3 < \ldots$ of H at scales $\gg 1/T$. For instance, from (2.4) we (formally) see that

$$\operatorname{tr}\left(\int_{\mathbf{R}} \eta(t/T) e^{itE_0} e^{-itH} \, dt\right) = T \sum_k \hat{\eta}\left(\frac{E_k - E_0}{2\pi/T}\right)$$

for any test function η and any energy level E_0. Roughly speaking, this formula tells us that the number of eigenvalues in an interval of size $O(1/T)$ can be more or less controlled by the Schrödinger operators up to time T.

A similar analysis also holds for the solution operator

$$u(t) = \cos(t\sqrt{-\Delta})u_0 + \frac{\sin(t\sqrt{-\Delta})}{\sqrt{-\Delta}} u_1$$

for the wave equation

$$\partial_t^2 u - \Delta u = 0$$

on an arbitrary spatial Riemannian manifold M (which we will take to be compact in order to have discrete spectrum). If we write λ_k for the eigenvalues of $\sqrt{-\Delta}$ (so the Laplace-Beltrami operator Δ has eigenvalues $-\lambda_k^2$), then a similar analysis to the above shows that knowledge of the solution to the wave equation up to time T gives (at least in principle) knowledge of the spectrum averaged to at the scale $1/T$ or above.

From the finite speed of propagation property of the wave equation (which has been normalised so that the speed of light c is equal to 1), one only needs to know the geometry of the manifold M up to distance scales T in order to understand the wave operator up to times T. In particular, if T is less than the injectivity radius of M, then the topology and global geometry of M is largely irrelevant, and the manifold more or less behaves like (a suitably normalised version of) Euclidean space. As a consequence, one can borrow Euclidean space techniques (such as the spatial Fourier transform) to control the spectrum at coarse scales $\gg 1$, leading in particular to the *Weyl law* for the distribution of eigenvalues on this manifold; see, for instance, [So1993] for a rigorous discussion. It is a significant challenge to go significantly below this scale and understand the finer structure of the spectrum; by the uncertainty principle, this task is largely equivalent to that of understanding the wave equation on long time scales $T \gg 1$, and the global geometry of the manifold M (and in particular, the dynamical properties of the geodesic flow) must then inevitably play a more dominant role.

Another important uncertainty principle duality relationship is that between the (imaginary parts of the) zeroes ρ of the Riemann zeta function $\zeta(s)$ and the logarithms $\log p$ of the primes. Starting from the fundamental

Euler product formula

$$\zeta(s) = \prod_p (1-p^{-s})^{-1}$$

and using rigorous versions of the heuristic factorisation

$$\zeta(s) \approx \prod_\rho (s-\rho)$$

one can soon derive explicit formulae connecting zeroes and primes, such as

$$\sum_\rho \frac{1}{s-\rho} \approx -\sum_p \log p \, e^{-s\log p}$$

(see e.g. [**Ta2010b**, §1.8] for more discussion). Using such formulae, one can relate the zeroes of the zeta function in the strip $\{\text{Im}(\rho) \leq T\}$ with the distribution of the log-primes at scales $\gg 1/T$. For instance, knowing that there are no zeroes on the line segment $\{1+it : |t| \leq T\}$ is basically equivalent to a partial prime number theorem $\pi(x) = (1+O(\frac{1}{T}))\frac{x}{\log x}$; letting $T \to \infty$, we see that the full prime number theorem is equivalent to the absence of zeroes on the entire line $\{1+it : t \in \mathbf{R}\}$. More generally, there is a fairly well-understood dictionary between the distribution of zeroes and the distribution of primes, which is explored in just about any advanced text in analytic number theory.

2.3.3. Phase space and the semi-classical limit. The above heuristic description of Fourier projections such as $1_{[-N,N]}(x)$ suggest that a Fourier projection $1_J(D)$ will approximately commute with a spatial projection $1_I(X)$ whenever I, J are intervals that obey the Heisenberg inequality $|I||J| \gg 1$. Again, this heuristic is not quite accurate if one uses sharp cutoffs (except in the dyadic model), but becomes quite valid if one uses smooth cutoffs. As such, one can morally talk about phase space projections $1_{I \times J}(X, D) \approx 1_I(X)1_J(D) \approx 1_J(D)1_I(X)$ to rectangles $I \times J$ in phase space, so long as these rectangles are large enough not to be in violation of the uncertainty principle.

Heuristically, $1_{I \times J}(X, D)$ is an orthogonal projection to the space[21] of functions that are localised to I in physical space and to J in frequency space. One can approximately compute the dimension of this not-quite-vector-space by computing the trace of the projection. Recalling that the trace of an integral operator $Tf(x) := \int_\mathbf{R} K(x,y)f(y) \, dy$ is given by $\text{tr}\, T =$

---

[21] This is morally a vector space, but unfortunately this is not rigorous due to the inability to perfectly localise in both physical space and frequency space simultaneously, thanks to the Hardy uncertainty principle.

2.3. The uncertainty principle

$\int_{\mathbf{R}} K(x,x)$, a short computation reveals that the trace of $1_I(X)1_J(D)$ is

$$\int_I \check{1}_J(0)\ dx = |I||J|.$$

Thus we conclude that the phase space region $I \times J$ contains approximately $|I||J|$ degrees of freedom in it, which can be viewed as a "macroscopic" version of the uncertainty principle.

More generally, the number of degrees of freedom contained in a large region $\Omega \subset \mathbf{R} \times \mathbf{R}$ of phase space is proportional to its area. Among other things, this can be used to justify the Weyl law for the distribution of eigenvalues of various operators. For instance, if H is the Schrödinger operator

$$H = -\hbar^2 \frac{d^2}{dx^2} + V(x) = \hbar^2 D^2 + V(X),$$

where $\hbar > 0$ is a small constant (which physically can be interpreted as Planck's constant), and V is a confining potential (to ensure discreteness of the spectrum), then the spectral projection $1_{[-\infty, E]}(H)$, when spectrally projected to energy levels below a given threshold E, is morally like a phase space projection to the region $\Omega := \{(\xi, x) : \hbar^2 \xi^2 + V(x) \leq E\}$. As such, the number of eigenvalues of H less than E should roughly equal the area of Ω, particularly when $\hbar$ is small (so that Ω becomes large, and the uncertainty principle no longer dominates); note that if V is a confining potential (such as the harmonic potential $V(x) = |x|^2$), then Ω will have finite area. Such heuristics can be justified by the machinery of *semi-classical analysis* and the *pseudo-differential calculus*, which we will not detail here.

The *correspondence principle* in quantum mechanics asserts that in the limit $\hbar \to 0$, quantum mechanics asymptotically converges (in some suitable sense) to classical mechanics. There are several ways to make this principle precise. One can work in a dual formulation, using algebras of observables rather than dealing with physical states directly, in which case the point is that the non-commutative operator algebras of quantum observables converge in various operator topologies to the commutative operator algebras of classical observables in the limit $\hbar \to 0$. This is the most common way that the correspondence principle is formulated; but one can also work directly using states. We illustrate this with the time-dependent Schrödinger equation

$$(2.5) \qquad i\hbar \partial_t \psi = -\frac{\hbar^2}{2m} \partial_{xx} \psi + V(x)\psi$$

with a potential V, where $m > 0$ is a fixed constant (representing mass) and $\hbar > 0$ is a small constant or, equivalently,

$$i\hbar \partial_t \psi = \left(\frac{P^2}{2m} + V(X)\right)\psi$$

where X is the position operator $Xf(x) := xf(x)$ and P is the momentum operator $Pf(x) := -i\hbar \frac{d}{dx} f(x)$ (thus $P = i\hbar D$). The classical counterpart to this equation is *Newton's second law*

$$F = ma$$

where $a = \frac{d^2 x}{dt^2}$ and $F = -\partial_x V(x)$; introducing the momentum $p := mv = m\frac{dx}{dt}$, one can rewrite Newton's second law as *Hamilton's equations of motion*

(2.6) $$\partial_t p = -\partial_x V(x), \quad \partial_t x = \frac{1}{m} p.$$

We now indicate (heuristically, at least) how (2.5) converges to (2.6) as $\hbar \to 0$. According to de Broglie's law $p = 2\pi \hbar \xi$, the momentum p should be proportional to the frequency ξ. Accordingly, consider a wave function ψ that at time t is concentrated near position $x_0(t)$ and momentum $p_0(t)$, and thus near frequency $p_0(t)/(2\pi\hbar)$; heuristically one can view ψ as having the shape

$$\psi(t,x) = A\left(t, \frac{x - x_0(t)}{r}\right) e^{ip_0(t)x/\hbar} e^{i\theta(t)/\hbar}$$

where $\theta(t)$ is some phase, r is some spatial scale (between 1 and $\hbar$) and A is some amplitude function. Informally, we have $X \approx x_0(t)$ and $P \approx p_0(t)$ for ψ.

Before we analyse the equation (2.5), we first look at some simpler equations. First, we look at

$$i\hbar \partial_t \psi = E\psi$$

where E is a real scalar constant. Then the evolution of this equation is given by a simple phase rotation:

$$\psi(t,x) = e^{-iEt/\hbar} \psi(0, x).$$

This phase rotation does not change the location $x_0(t)$ or momentum $p_0(t)$ of the wave:

$$\partial_t x_0(t) = 0, \quad \partial_t p_0(t) = 0.$$

Next, we look at the transport equation

$$i\hbar \partial_t \psi = -i\hbar v \partial_x \psi$$

where $v \in \mathbf{R}$ is another constant This evolution is given by translation

$$\psi(t,x) = \psi(0, x - vt);$$

the position $x_0(t)$ of this evolution moves at the constant speed of v, but the momentum is unchanged:

$$\partial_t x_0(t) = v; \quad \partial_t p_0(t) = 0.$$

Combining the two, we see that an equation of the form

$$i\hbar \partial_t \psi = E\psi - i\hbar v (\partial_x - ip_0(t)/\hbar)\psi$$

2.3. The uncertainty principle

would also transport the position x_0 at a constant speed of v, without changing the momentum. Next, we consider the modulation equation

$$i\hbar\partial_t\psi = Fx\psi$$

where $F \in \mathbf{R}$ is yet another constant. This equation is solved by the formula

$$\psi(t,x) = e^{itFx/\hbar}\psi(0,x);$$

this phase modulation does not change the position $x_0(t)$, but steadily increases the momentum $p_0(t)$ at a rate of F:

$$\partial_t x_0(t) = 0, \quad \partial_t p_0(t) = F.$$

Finally, we combine all these equations together, looking at the combined equation

$$i\hbar\partial_t\psi = E\psi - i\hbar v(\partial_x - ip_0(t)/\hbar)\psi + F(x - x_0(t))\psi.$$

Heuristically at least, the position $x_0(t)$ and momentum $p_0(t)$ of solutions to this equation should evolve according to the law

(2.7) $$\partial_t x_0(t) = v, \quad \partial_t p_0(t) = F.$$

Remark 2.3.5. One can make the above discussion more rigorous by using the *metaplectic representation*.

The above analysis was for v, F constant, but as all statements here are instantaneous and first-order in time, it also applies for time-dependent v, F.

Now we return to the Schrödinger equation (2.5). If ψ is localised in space close to $x_0(t)$, then by Taylor expansion we may linearise the $V(x)$ component as

$$V(x) = V(x_0(t)) + (x - x_0(t))\partial_x V(x).$$

Similarly, if ψ is localised in momentum close to $p_0(t)$, then in frequency it is localised close to $p_0(t)/(2\pi\hbar)$, so that $\partial_x \approx ip_0(t)/\hbar$, and so we have a Taylor expansion

$$\partial_{xx} \approx (ip_0(t)/\hbar)^2 + 2(ip_0(t)/\hbar)\left(\partial_x - (ip_0(t)/\hbar)\right).$$

These Taylor expansions become increasingly accurate in the limit $\hbar \to 0$, assuming suitable localisation in both space and momentum. Inserting these approximations and simplifying, one arrives at

$$\partial_t\psi = \frac{E(t)}{i\hbar}\psi - \frac{p_0(t)}{m}\left(\partial_x - (ip_0(t)/\hbar)\right)\psi - \frac{i}{\hbar}(x - x_0(t))\partial_x V(x_0(t))\psi$$

where $E(t) := \frac{p_0(t)^2}{2m} + V(x_0(t))$ is the classical energy of the state. Using the heuristics (2.7) we are led to (2.6) as desired.

More generally, a Schrödinger equation

$$i\hbar\partial_t\psi = H(X, P)\psi$$

where $P := -i\hbar \frac{d}{dx}$ is the momentum operator, and being vague about exactly what a function $H(X, P)$ of two non-commuting operators X, P means, can be (heuristically) approximately Taylor expanded as

$$i\hbar \partial_t \psi = H(x_0(t), p_0(t))$$
$$+ \frac{\partial H}{\partial p} H(x_0(t), p_0(t))(P - p_0(t))\psi$$
$$+ \frac{\partial H}{\partial x} H(x_0(t), p_0(t))(X - x_0(t))\psi$$

and (2.7) leads us to the Hamilton equations of motion

$$\partial_t x(t) = \frac{\partial H}{\partial p}, \quad \partial_t p(t) = -\frac{\partial H}{\partial x}.$$

It turns out that these heuristic computations can be made completely rigorous in the semi-classical limit $\hbar \to 0$, by using the machinery of *pseudo-differential calculus*, but we will not go into detail here.

Bibliography

[AlBe2001] N. Alon, R. Beigel, *Lower bounds for approximations by low degree polynomials over $\mathbf{Z}_m$*, Proc. of the 16th Annual IEEE Conference on Computational Complexity (CCC), IEEE, 2001, pp. 184-187.

[AlKaKrLiRo2003] N. Alon, T. Kaufman, M. Krivelevich, S. Litsyn, D. Ron, *Testing low-degree polynomials over GF(2)*, Approximation, randomization, and combinatorial optimization, 188-199, Lecture Notes in Comput. Sci., 2764, Springer, Berlin, 2003.

[BaKa2011] M. Bateman, N. Katz, *New Bounds on cap sets*, preprint. `arXiv:1101.5851`

[Be1946] F. A. Behrend, *On sets of integers which contain no three terms in arithmetical progression*, Proc. Nat. Acad. Sci. U. S. A. **32** (1946), 331-332.

[BeCaChTa2008] J. Bennett, Jonathan; A. Carbery, M. Christ, T. Tao, *The Brascamp-Lieb inequalities: finiteness, structure and extremals*, Geom. Funct. Anal. **17** (2008), no. 5, 1343-1415.

[BeHoKa2005] V. Bergelson, B. Host and B. Kra, *Multiple recurrence and nilsequences*, with an appendix by Imre Ruzsa, Invent. Math. **160** (2005), no. 2, 261-303.

[BeTaZi2010] V. Bergelson, T. Tao, T. Ziegler, *An inverse theorem for the uniformity seminorms associated with the action of F_p*, Geom. Funct. Anal. **19** (2010), no. 6, 1539-1596.

[BlLuRu1993] M. Blum, M. Luby, R. Rubinfeld, *Self-testing/correcting with applications to numerical problems*, Proceedings of the 22nd Annual ACM Symposium on Theory of Computing (Baltimore, MD, 1990). J. Comput. System Sci. **47** (1993), no. 3, 549-595.

[BoVi2010] A. Bogdanov, E. Viola, *Pseudorandom bits for polynomials*, SIAM J. Comput. **39** (2010), no. 6, 2464-2486.

[Bo1986] J. Bourgain, *A Szemerédi type theorem for sets of positive density in $\mathbf{R}^k$*, Israel J. Math. **54** (1986), no. 3, 307-316.

[Bo1999] J. Bourgain, *On triples in arithmetic progression*, Geom. Funct. Anal. **9** (1999), no. 5, 968-984.

[Bo2008] J. Bourgain, *Roth's theorem on progressions revisited*, J. Anal. Math. **104** (2008), 155-192.

[BrGrTa2010] E. Breuillard, B. Green, T. Tao, *Approximate subgroups of linear groups*, Geom. Funct. Anal. **21** (2011), no. 4, 774-819.

[CaSz2010] O. Camarena, B. Szegedy, *Nilspaces, nilmanifolds and their morphisms*, preprint. arXiv:1009.3825

[CoLe1984] J.-P. Conze, E. Lesigne, *Théorémes ergodiques pour des mesures diagonales*, Bull. Soc. Math. France **112** (1984), 143-175.

[El2008] M. Elkin, *An improved construction of progression-free sets*, Israel J. Math. **184** (2011), 93-128.

[Fu1977] H. Furstenberg, *Ergodic behavior of diagonal measures and a theorem of Szemerédi on arithmetic progressions*, J. Analyse Math. **31** (1977), 204-256.

[Fu1990] H. Furstenberg, *Nonconventional ergodic averages*, The legacy of John von Neumann (Hempstead, NY, 1988), 43-56, Proc. Sympos. Pure Math., 50, Amer. Math. Soc., Providence, RI, 1990.

[FuWi1996] H. Furstenberg, B. Weiss, *A mean ergodic theorem for $1/N \sum_{n=1}^{N} f(T^n x) g(T^{n^2} x)$*. Convergence in ergodic theory and probability (Columbus, OH, 1993), 193-227, Ohio State Univ. Math. Res. Inst. Publ., 5 de Gruyter, Berlin, 1996.

[GoPiYi2008] D. Goldston, J. Pintz, C. Yıldırım, *Primes in Tuples II*, Acta Math. **204** (2010), no. 1, 1-47.

[Go1998] W. T. Gowers, *A new proof of Szemerédi's theorem for arithmetic progressions of length four*, Geom. Funct. Anal. **8** (1998), no. 3, 529-551.

[Go2001] W. T. Gowers, *A new proof of Szemerédi's theorem*, Geom. Funct. Anal. **11** (2001), no. 3, 465-588.

[Go2010] W. T. Gowers, *Decompositions, approximate structure, transference, and the Hahn-Banach theorem*, Bull. Lond. Math. Soc. **42** (2010), no. 4, 573-606.

[GoWo2010] W. T. Gowers, J. Wolf, *The true complexity of a system of linear equations*, Proc. Lond. Math. Soc. (3) **100** (2010), no. 1, 155-176.

[GoWo2010b] W. T. Gowers, J. Wolf, *Linear forms and higher-degree uniformity for functions on $\mathbb{F}_p^n$*, Geom. Funct. Anal. **21** (2011), no. 1, 36-69.

[GrRoSp1980] R. Graham, B. Rothschild, J.H. Spencer, Ramsey Theory, *John Wiley and Sons*, NY (1980).

[Gr2005] B. Green, *Roth's theorem in the primes*, Annals of Math. **161** (2005), no. 3, 1609-1636.

[Gr2005b] B. Green, *A Szemerédi-type regularity lemma in abelian groups, with applications*, Geom. Funct. Anal. **15** (2005), no. 2, 340-376.

[Gr2005a] B. Green, *Finite field models in additive combinatorics*, Surveys in combinatorics 2005, 127, London Math. Soc. Lecture Note Ser., 327, Cambridge Univ. Press, Cambridge, 2005.

[Gr2007] B. Green, *Montréal lecture notes on quadratic Fourier analysis*, Additive Combinatorics (Montréal 2006, ed. Granville et al.), CRM Proceedings vol. 43, 69-102, AMS 2007.

[GrKo2009] B. Green, S. Konyagin, *On the Littlewood problem modulo a prime*, Canad. J. Math. **61** (2009), no. 1, 141-164.

[GrTa2006] B. Green, T. Tao, *Restriction theory of the Selberg sieve, with applications*, J. Th. Nombres Bordeaux **18** (2006), 137-172.

[GrTa2008] B. Green, T. Tao, *An inverse theorem for the Gowers $U^3(G)$ norm*, Proc. Edinb. Math. Soc. (2) **51** (2008), no. 1, 73-153.

[GrTa2008b] B. Green, T. Tao, *The primes contain arbitrarily long arithmetic progressions*, Annals of Math. **167** (2008), 481-547.

[GrTa2008c] B. Green, T. Tao, *The Möbius function is strongly orthogonal to nilsequences*, preprint. arxiv:0807.1736

[GrTa2009] B. Green, T. Tao, *The distribution of polynomials over finite fields, with applications to the Gowers norms*, Contrib. Discrete Math. **4** (2009), no. 2, 1-36.

[GrTa2010] B. Green, T. Tao, *Linear equations in primes*, Annals of Math. **171** (2010), 1753-1850.

[GrTa2010b] B. Green, T. Tao, *An arithmetic regularity lemma, an associated counting lemma, and applications*, An Irregular Mind: Szemeredi is 70, Bolyai Society Mathematical Studies, 2010.

[GrTa2011] B. Green, T. Tao, *The quantitative behaviour of polynomial orbits on nilmanifolds*, preprint. arXiv:0709.3562

[GrTaZi2009] B. Green, T. Tao, T. Ziegler, *An inverse theorem for the Gowers $U^4[N]$ norm*, preprint. arXiv:0911.5681

[GrTaZi2010] B. Green, T. Tao, T. Ziegler, *An inverse theorem for the Gowers $U^{s+1}[N]$-norm (Announcement)*, Electron. Res. Announc. Math. Sci. **18** (2011), 69-90.

[GrTaZi2010b] B. Green, T. Tao, T. Ziegler, *An inverse theorem for the Gowers $U^{s+1}[N]$-norm*, preprint. arXiv:1009.3998

[GrWo2008] B. Green, J. Wolf, *A note on Elkin's improvement of Behrend's construction*, Additive number theory, 141-144, Springer, New York, 2010.

[Gr1961] L. W. Green, *Spectra of nilflows*, Bull. Amer. Math. Soc. **67** (1961) 414-415.

[Gr1981] M. Gromov, *Groups of polynomial growth and expanding maps*, Inst. Hautes Études Sci. Publ. Math. No. **53** (1981), 53-73.

[Ha1993] I. J. Håland, *Uniform distribution of generalized polynomials*, J. Number Theory **45** (1993), 327-366.

[HaSh2010] E. Haramaty, A. Shpilka, *On the structure of cubic and quartic polynomials*, STOC'10—Proceedings of the 2010 ACM International Symposium on Theory of Computing, 331-340, ACM, New York, 2010.

[HaLo2010] H. Hatami, S. Lovett, *Higher-order Fourier analysis of $\mathbb{F}_p^n$ and the complexity of systems of linear forms*, Geom. Funct. Anal. **21** (2011), no. 6, 1331-1357.

[HB1987] D. R. Heath-Brown, *Integer sets containing no arithmetic progressions*, J. London Math. Soc. (2) **35** (1987), no. 3, 385-394.

[Ho2006] B. Host, *Progressions arithmétiques dans les nombres premiers (d'aprés B. Green et T. Tao)* Séminaire Bourbaki. Vol. 2004/2005. Astérisque, No. 307 (2006), Exp. No. 944, viii, 229-246.

[HoKr2005] B. Host, B. Kra, *Nonconventional ergodic averages and nilmanifolds*, Ann. of Math. (2) **161** (2005), no. 1, 397-488.

[IwKo2004] H. Iwaniec, E. Kowalski, Analytic number theory. American Mathematical Society Colloquium Publications, 53. American Mathematical Society, Providence, RI, 2004.

[KaLo2008] T. Kaufman, S. Lovett, *Worst case to average case reductions for polynomials*, FOCS (2008), 166-175.

[Kl1971] S. L. Kleiman, *Les théorèmes de finitude pour le foncteur de Picard*, in *Théorie des intersections et théorème de Riemann-Roch* (SGA6), exposé XIII, pp. 616-666. LNM 225, Springer-Verlag, Berlin–New York, 1971.

[KoLuRo1996] Y. Kohayakawa, T. Luczsak, V. Rödl, *Arithmetic progressions of length three in subsets of a random set*, Acta Arith. **75** (1996), no. 2, 133-163.

[Kr2006] B. Kra, *From combinatorics to ergodic theory and back again*, International Congress of Mathematicians. Vol. III, 5776, Eur. Math. Soc., Zürich, 2006.

[Kr2007] B. Kra, *Ergodic methods in additive combinatorics*, Additive combinatorics, 103-143, CRM Proc. Lecture Notes, 43, Amer. Math. Soc., Providence, RI, 2007.

[La1954] M. Lazard, *Sur les groupes nilpotents et les anneaux de Lie*, Ann. Sci. Ecole Norm. Sup. (3) **71** (1954), 101-190

[Le1998] A. Leibman, *Polynomial sequences in groups*, Journal of Algebra 201 (1998), 189-206.

[Le2002] A. Leibman, *Polynomial mappings of groups*, Israel J. Math. **129** (2002), 29-60.

[Le2005] A. Leibman, *Pointwise convergence of ergodic averages of polynomial sequences of translations on a nilmanifold*, Ergodic Theory and Dynamical Systems **25** (2005), no. 1, 201-213.

[Lo1975] P. A. Loeb, *Conversion from nonstandard to standard measure spaces and applications in probability theory*, Trans. Amer. Math. Soc. **211** (1975), pp. 113-122.

[LoMeSa2008] S. Lovett, R. Meshulam, A. Samorodnitsky, *Inverse conjecture for the Gowers norm is false*, STOC'08, 547-556, ACM, New York, 2008.

[Ma1949] A. Mal'cev, *On a class of homogeneous spaces*, Izvestiya Akad. Nauk SSSR, Ser Mat. **13** (1949), 9-32.

[Me1995] R. Meshulam, *On subsets of finite abelian groups with no 3-term arithmetic progressions*, J. Combin. Theory Ser. A **71** (1995), no. 1, 168-172.

[Mu1970] D. Mumford, *Varieties defined by quadratic equations*, 1970 Questions on Algebraic Varieties (C.I.M.E., III Ciclo, Varenna, 1969) pp. 29-100 Edizioni Cremonese, Rome.

[Pa1970] W. Parry, *Dynamical systems on nilmanifolds*, Bull. London Math. Soc. **2** (1970) 37-40.

[ReTrTuVa2008] O. Reingold, L. Trevisan, M. Tulsiani, S. Vadhan, *New Proofs of the Green-Tao-Ziegler Dense Model Theorem: An Exposition*, preprint. arXiv:0806.0381

[Ro1953] K.F. Roth, *On certain sets of integers*, J. London Math. Soc. **28** (1953), 245-252.

[Sa2007] A. Samorodnitsky, *Low-degree tests at large distances*, STOC'07 in Proceedings of the 39th Annual ACM Symposium on Theory of Computing, 506515, ACM, New York, 2007.

[Sa2010] T. Sanders, *On certain other sets of integers*, J. Anal. Math., to appear, arXiv:1007.5444, 2010.

[Sa2010] T. Sanders, *On Roth's theorem on progressions*, preprint.

[Sh2009] Y. Shalom, T. Tao, *A finitary version of Gromov's polynomial growth theorem*, Geom. Funct. Anal. 20 (2010), no. 6, 1502-1547.

[So1993] C. Sogge, Fourier integrals in classical analysis. Cambridge Tracts in Mathematics, 105, Cambridge University Press, Cambridge, 1993.

[SuTrVa1999] M. Sudan, L. Trevisan, S. Vadhan, *Pseudorandom generators without the XOR lemma*, Annual ACM Symposium on Theory of Computing (Atlanta, GA, 1999), 537546 (electronic), ACM, New York, 1999.

[Sz2009] B. Szegedy, *Higher order Fourier analysis as an algebraic theory I*, preprint. arXiv:0903.0897

[Sz2009b] B. Szegedy, *Higher order Fourier analysis as an algebraic theory II*, preprint. arXiv:0911.1157

[Sz2010] B. Szegedy, *Higher order Fourier analysis as an algebraic theory III*, preprint. arXiv:1001.4282

[Sz2010b] B. Szegedy, *Gowers norms, regularization and limits of functions on abelian groups*, preprint. arXiv:1010.6211

[Sz2010c] B. Szegedy, *Structure of finite nilspaces and inverse theorems for the Gowers norms in bounded exponent groups*, preprint. arXiv:1011.1057

[Sz1975] E. Szemerédi, *On sets of integers containing no k elements in arithmetic progression*, Acta Arith. **27** (1975), 299-345.

[Sz1978] E. Szemerédi, *Regular partitions of graphs*, in "Problemés Combinatoires et Théorie des Graphes, Proc. Colloque Inter. CNRS," (Bermond, Fournier, Las Vergnas, Sotteau, eds.), CNRS Paris, 1978, 399-401.

[Ta2003] T. Tao, *Recent progress on the Restriction conjecture*, preprint. math.CA/0311181

[Ta2004] T. Tao, *A remark on Goldston-Yıldırım correlation estimates*, unpublished. www.math.ucla.edu/~tao/preprints/Expository/gy-corr.dvi

[Ta2006] T. Tao, *Szemerédi's regularity lemma revisited*, Contrib. Discrete Math. **1** (2006), no. 1, 8-28.

[Ta2006b] T. Tao, Nonlinear dispersive equations: local and global analysis, CBMS regional series in mathematics, 2006.

[Ta2007] T. Tao, *Structure and randomness in combinatorics*, Proceedings of the 48th Annual Symposium on Foundations of Computer Science (FOCS) 2007, 3-18.

[Ta2008] T. Tao, Structure and Randomness, American Mathematical Society, 2008.

[Ta2009] T. Tao, Poincaré's legacies, Vol. I., American Mathematical Society, 2009.

[Ta2010] T. Tao, An epsilon of room, Vol. I., Graduate Studies in Mathematics, 117. American Mathematical Society, Providence, RI, 2010.

[Ta2010b] T. Tao, An epsilon of room, Vol. II., American Mathematical Society, 2010.

[Ta2011] T. Tao, An introduction to measure theory, American Mathematical Society, 2011.

[Ta2011b] T. Tao, Topics in random matrix theory, American Mathematical Society, 2011.

[TaVu2006] T. Tao, V. Vu, Additive combinatorics. Cambridge Studies in Advanced Mathematics, 105, Cambridge University Press, Cambridge, 2006.

[TaZi2008] T. Tao, T. Ziegler, *The primes contain arbitrarily long polynomial progressions*, Acta Math. **201** (2008), 213-305.

[TaZi2010] T. Tao, T. Ziegler, *The inverse conjecture for the Gowers norm over finite fields via the correspondence principle*, Anal. PDE **3** (2010), no. 1, 1-20.

[TaZi2011] T. Tao, T. Ziegler, *The inverse conjecture for the Gowers norm over finite fields in low characteristic*, preprint. arXiv:1101.1469

[Wo] J. Wolf, *The minimum number of monochromatic 4-term progressions*, www.juliawolf.org/research/preprints/talk280509.pdf

[Zi2007] T. Ziegler, *Universal characteristic factors and Furstenberg averages*, J. Amer. Math. Soc. 20 (2007), no. 1, 53-97.

Index

2-coboundary, 78
99% inverse theorem for the Gowers norms, 75
W-trick, 124
δ-equidistribution, 12

additive cohomology, 77
additive quadruple, 83
algebraic set, 138
algebraic variety, 142
almost periodicity, 40
analytic rank, 72
arithmetic regularity lemma (strong), 38
arithmetic regularity lemma (weak), 38
asymptotic equidistribution, 3, 10
asymptotic notation, 3
asymptotic notation (ultralimit analysis), 21
atom, 33

Balog-Szemerédi-Gowers-Freiman theorem, 84
Bezout's theorem, 141
bias, 64
Bogdanov-Viola lemma, 66
Bohr set, 116
bounded (ultralimit analysis), 21
bracket polynomial, 102

Cauchy-Schwarz complexity, 55
Cauchy-Schwarz inequality, 149
Cauchy-Schwarz-Gowers inequality, 58, 157

characteristic, 55
Chevalley-Warning theorem, 63
classical polynomial, 61
cocycle, 77
complex conjugation, 151
complexity of a nilmanifold, 101
complexity of a nilsequence, 102
complexity of an algebraic set, 139
conditional expectation, 33
continuity of dimension, 140
continuity of irreducibility, 143
converse inverse theorem for the Gowers norms, 75
correlation condition, 121
correspondence principle, 175

de Broglie's law, 176
degree, 94
dense model theorem, 111, 121
density increment argument, 28
differentiation of nilsequences, 104
dimension, 140
Dirac measure, 3

energy increment argument, 33
equidistribution, 62
Equidistribution (abelian linear sequences), 7
Equidistribution (abelian multidimensional polynomial sequences), 11
Equidistribution (abelian polynomial sequences), 9, 17, 24

equidistribution (ultralimit analysis), 22
equidistribution theorem, 6
error correction of polynomials, 67
exponential sum, 2

factor, 33
Fejér summation, 13
filtered group, 94
filtration, 94
Fourier measurability, 35
Fourier pseudorandomness, 48
fragmentation, 31

generalised von Neumann inequality, 59
generalised von Neumann theorem, 118
Gowers box space, 154
Gowers inner product, 57, 154
Gowers triangle inequality, 157
Gowers uniformity norm, 47, 57, 154
Gowers' Cauchy-Schwarz argument, 83
Gowers-Host-Kra semi-norm, 155
Gowers-Wolf theorem, 92
Gromov's theorem, 145
growth spurt, 19

Haar measure, 5
Hall-Petresco formula, 98
Hamilton's equation of motion, 176
Hardy-Littlewood maximal inequality, 36
Heisenberg group, 98
Heisenberg nilmanifold, 100
Higher order inner product space, 153
Hilbert cube lemma, 46
Hilbert space, 150
horizontal character, 108
horizontal torus, 108
Host-Kra group, 93, 94
Host-Kra measure, 155
hyperreal, 135

indicator function, 2
infinitesimal, 21, 138
inner product space, 149
inverse conjecture for the Gowers norm, 58, 106
inverse conjecture for the Gowers uniformity norms, 75
irrational, 6

join, 34

Kronecker factor, 42

Kronecker measurability, 42

Lazard-Leibman theorem, 96
Liebman equidistribution criterion, 108
limit finite set, 20
limit function, 21
limit number, 20
limit object, 134
limit set, 20
linear phase, 2
linear forms conditions, 118
Lipschitz norm, 11
Littlewood-Paley theory, 171
local Gowers inner product, 160
local testability, 79
Loeb measure, 41
Los's theorem, 20
low rank, 65
lower central series, 94
lower face, 95
Lucas' theorem, 80

Mal'cev basis, 99
multiple recurrence, 19

Newton's second law, 176
nilpotent, 98
Noetherian condition, 142
non-principal ultrafilter, 133
norm, 150

phase heuristic, 167
polyar body, 168
polynomial, 54, 60
polynomial orbit, 101
polynomial phase invariance, 59
polynomial recurrence, 9
polynomial sequence, 98
Pontryagin dual, 57

Ramsey's theorem, 80
recurrence, 19
refinement, 34
regularity lemma for polynomials, 71
relative van der Corput lemma, 107
reproducing formula, 170
restriction estimate, 114
rigidity, 67
Roth's theorem, 26, 33, 112
Roth's theorem in the primes, 110
Roth-pseudorandom, 112

Schrödinger equation, 172, 175

Index

semi-norm, 150
splitting axiom, 153
standard, 20
standard part, 21, 138
standard universe, 131
strong arithmetic regularity lemma, 91
structure and randomness, 34
superstructure, 131
symmetric polynomial, 80
syndeticity, 19

Taylor coefficient, 98
tensor product, 151
total asymptotic equidistribution, 4, 10
total equidistribution (ultralimit analysis), 22
transference, 110
triangle inequality, 149

ultralimit, 20, 21, 134
ultrapower, 20, 134
ultraproduct, 20, 134

van der Corput inequality, 7
van der Corput lemma, 8, 11, 24, 65
vertical character, 103
vertical frequency, 103
Vinogradov lemma, 15, 24
von Mangoldt function, 124

weight function, 112
Weyl criterion, 13
Weyl equidistribution criterion, 5, 10, 13, 23, 64
Weyl equidistribution theorem, 9
Weyl law, 173

图字：01-2016-2502 号

Higher Order Fourier Analysis, by Terence Tao, first published by the American Mathematical Society.
Copyright © 2012 Terence Tao. All rights reserved.
This present reprint edition is published by Higher Education Press Limited Company under authority
of the American Mathematical Society and is published under license.
Special Edition for People's Republic of China Distribution Only. This edition has been authorized by
the American Mathematical Society for sale in People's Republic of China only, and is not for export therefrom.

本书最初由美国数学会于 2012 年出版，原书名为 *Higher Order Fourier Analysis*，作者为 Terence Tao。
Terence Tao 保留原书所有版权。
原书版权声明：Copyright © 2012 by Terence Tao。
本影印版由高等教育出版社有限公司经美国数学会独家授权出版。
本版只限于中华人民共和国境内发行。本版经由美国数学会授权仅在中华人民共和国境内销售，不得出口。

高阶傅里叶分析
Gaojie Fuliye Fenxi

图书在版编目 (CIP) 数据

高阶傅里叶分析 = Higher Order Fourier Analysis：
英文 /（澳）陶哲轩 (Terence Tao) 著. —影印本.
—北京：高等教育出版社，2017.1
ISBN 978-7-04-046909-7

Ⅰ. ①高… Ⅱ. ①陶… Ⅲ. ①傅里叶分析—英文
Ⅳ. ①O174.2

中国版本图书馆 CIP 数据核字 (2016) 第 281867 号

| 策划编辑 | 赵天夫 | 责任编辑 | 赵天夫 |
| 封面设计 | 张申申 | 责任印制 | 毛斯璐 |

出版发行	高等教育出版社	开本	787mm×1092mm 1/16
社址	北京市西城区德外大街 4 号	印张	12.75
邮政编码	100120	字数	263 千字
购书热线	010-58581118	版次	2017 年 1 月第 1 版
咨询电话	400-810-0598	印次	2017 年 1 月第 1 次印刷
网址	http://www.hep.edu.cn	定价	99.00 元
	http://www.hep.com.cn		
网上订购	http://www.hepmall.com.cn	本书如有缺页、倒页、脱页等质量问题，
	http://www.hepmall.com	请到所购图书销售部门联系调换
	http://www.hepmall.cn	版权所有　侵权必究
印刷	北京中科印刷有限公司	[物 料 号 46909-00]

郑重声明

高等教育出版社依法对本书享有专有出版权。任何未经许可的复制、销售行为均违反《中华人民共和国著作权法》，其行为人将承担相应的民事责任和行政责任；构成犯罪的，将被依法追究刑事责任。为了维护市场秩序，保护读者的合法权益，避免读者误用盗版书造成不良后果，我社将配合行政执法部门和司法机关对违法犯罪的单位和个人进行严厉打击。社会各界人士如发现上述侵权行为，希望及时举报，本社将奖励举报有功人员。

反盗版举报电话　(010) 58581999　58582371　58582488
反盗版举报传真　(010) 82086060
反盗版举报邮箱　dd@hep.com.cn
通信地址　北京市西城区德外大街 4 号
　　　　　高等教育出版社法律事务与版权管理部
邮政编码　100120

美国数学会经典影印系列

1. **Lars V. Ahlfors**, Lectures on Quasiconformal Mappings, Second Edition, 978-7-04-047010-9
2. **Dmitri Burago, Yuri Burago, Sergei Ivanov**, A Course in Metric Geometry, 978-7-04-046908-0
3. **Tobias Holck Colding, William P. Minicozzi II**, A Course in Minimal Surfaces, 978-7-04-046911-0
4. **Javier Duoandikoetxea**, Fourier Analysis, 978-7-04-046901-1
5. **John P. D'Angelo**, An Introduction to Complex Analysis and Geometry, 978-7-04-046998-1
6. **Y. Eliashberg, N. Mishachev**, Introduction to the h-Principle, 978-7-04-046902-8
7. **Lawrence C. Evans**, Partial Differential Equations, Second Edition, 978-7-04-046935-6
8. **Robert E. Greene, Steven G. Krantz**, Function Theory of One Complex Variable, Third Edition, 978-7-04-046907-3
9. **Thomas A. Ivey, J. M. Landsberg**, Cartan for Beginners: Differential Geometry via Moving Frames and Exterior Differential Systems, 978-7-04-046917-2
10. **Jens Carsten Jantzen**, Representations of Algebraic Groups, Second Edition, 978-7-04-047008-6
11. **A. A. Kirillov**, Lectures on the Orbit Method, 978-7-04-046910-3
12. **Jean-Marie De Koninck, Armel Mercier**, 1001 Problems in Classical Number Theory, 978-7-04-046999-8
13. **Peter D. Lax, Lawrence Zalcman**, Complex Proofs of Real Theorems, 978-7-04-047000-0
14. **David A. Levin, Yuval Peres, Elizabeth L. Wilmer**, Markov Chains and Mixing Times, 978-7-04-046994-3
15. **Dusa McDuff, Dietmar Salamon**, J-holomorphic Curves and Symplectic Topology, 978-7-04-046993-6
16. **John von Neumann**, Invariant Measures, 978-7-04-046997-4
17. **R. Clark Robinson**, An Introduction to Dynamical Systems: Continuous and Discrete, Second Edition, 978-7-04-047009-3
18. **Terence Tao**, An Epsilon of Room, I: Real Analysis: pages from year three of a mathematical blog, 978-7-04-046900-4
19. **Terence Tao**, An Epsilon of Room, II: pages from year three of a mathematical blog, 978-7-04-046899-1
20. **Terence Tao**, An Introduction to Measure Theory, 978-7-04-046905-9
21. **Terence Tao**, Higher Order Fourier Analysis, 978-7-04-046909-7
22. **Terence Tao**, Poincaré's Legacies, Part I: pages from year two of a mathematical blog, 978-7-04-046995-0
23. **Terence Tao**, Poincaré's Legacies, Part II: pages from year two of a mathematical blog, 978-7-04-046996-7
24. **Cédric Villani**, Topics in Optimal Transportation, 978-7-04-046921-9
25. **R. J. Williams**, Introduction to the Mathematics of Finance, 978-7-04-046912-7